Applied
Underwater Acoustics

BY

D. G. TUCKER, D.Sc.

*Professor and Head of Department of Electronic and Electrical
Engineering, University of Birmingham*

AND

B. K. GAZEY, Ph.D.

*Lecturer, Department of Electronic and Electrical Engineering,
University of Birmingham*

PERGAMON PRESS

OXFORD · NEW YORK · TORONTO · SYDNEY
PARIS · FRANKFURT

U.K.	Pergamon Press Ltd., Headington Hill Hall, Oxford OX3 0BW, England
U.S.A.	Pergamon Press Inc., Maxwell House, Fairview Park, Elmsford, New York 10523, U.S.A.
CANADA	Pergamon of Canada Ltd., 75 The East Mall, Toronto, Ontario, Canada
AUSTRALIA	Pergamon Press (Aust.) Pty. Ltd., 19a Boundary Street, Rushcutters Bay, N.S.W. 2011, Australia
FRANCE	Pergamon Press SARL, 24 rue des Ecoles, 75240 Paris, Cedex 05, France
FEDERAL REPUBLIC OF GERMANY	Pergamon Press GmbH, 6242 Kronberg-Taunus, Pferdstrasse 1, Federal Republic of Germany

First edition 1966

Reprinted with corrections 1977

Library of Congress Catalog Card No. 66–18403

Printed in Great Britain by Biddles Ltd., Guildford, Surrey

ISBN 0 08 011816 X flexicover

D
620.25
TUC

Contents

Preface

WE HAVE written this book with the needs of university and college students in mind as well as those of people starting on a career in underwater research and engineering. With the rapid growth in oceanography, sonar, underwater telemetry and communication, marine geophysics, and other fields in which underwater acoustics play an important role, there seems a need for a basic textbook which combines the physicist's and the engineer's approach. After a long period of decline, applied acoustics is now finding its way back into university syllabuses, and it is to be hoped that suitable emphasis will be given to underwater problems in such academic courses. The book is based on our own courses in the Department of Electronic and Electrical Engineering at the University of Birmingham, where at final-year undergraduate level a forty-lecture course on Applied Acoustics is offered which comprises the work in this book, together with auditory acoustics, and at postgraduate level a twenty-lecture course is given. As we have a large research programme in this field at Birmingham, the choice of material for the book may have been influenced by what we think is going to be important in the future rather than by what has hitherto been important. But we would like to reiterate that this is a textbook, based firmly on classical acoustics, and is in no sense a research book.

Birmingham   D. G. T.
                                                               B. K. G.

List of Symbols

ϕ_r	Phase angle between two adjacent receiving channels
	Random sequence of phase differences
$\phi(t)$	Phase angle varying randomly with time
ψ	Correlation coefficient
ω_0	Midband angular frequency
$\omega_p, \omega_q, \omega_r$	Angular frequency of noise components p, q and r

CHAPTER 3

A	Amplitude
	Area
	A constant
a	Particle acceleration. Absorption coefficient (dB/m)
a_1	Radius of target sphere
B	A constant
C_p	Specific heat at constant pressure
C_v	Specific heat at constant volume
c	Propagation velocity
c_{AD}	Propagation velocity under adiabatic conditions
c_{IS}	Propagation velocity under isothermal conditions
D	Recognition differential
D.F.	Directivity factor
D.I.	Directivity index
d	Peak value of particle displacement
E	Instantaneous energy
f	Frequency
f_r	Relaxation frequency
H	Transmission anomaly
I	Acoustic intensity
I_E	Echo intensity
I_I	Incident acoustic intensity
I_N	Effective noise intensity level
I_0	Acoustic intensity at index point
I_R	Acoustic intensity level at range r
I_S	Source intensity level
I_s	Scattered acoustic intensity
K	Thermal conductivity
K.E.	Kinetic energy
k	Acoustic wave number $= 2\pi/\lambda$
l	Distance
M	Source strength
N	Propagation loss
N.P.	Noise power
n	An integer
	A non-integral exponent
P	Instantaneous pressure
P.E.	Potential energy
P_0	Hydrostatic pressure
p	Acoustic pressure

p_0	Mean value of p
R	Specific acoustic resistance
	Reverberation intensity level
R_I	Reverberation intensity
r	Radius
	Radial distance
r, θ, ϕ	Spherical co-ordinates
r_0	Radius of spherical acoustic source
S	Condensation
s	Compressibility
T	Target strength
t	Time
u	Particle velocity
u, v, w	Components of velocity
u_0	Surface velocity of spherical acoustic source
v	Volume
v_0	Initial volume
W	Radiated acoustic power
X	Specific acoustic reactance
x, y, z	Rectangular co-ordinates'
Z	Specific acoustic impedance
Z_0	Characteristic impedance
	Characteristic acoustic impedance ($\rho_0 c_0$)
Z_S	Source impedance
	Source acoustic impedance ($\rho_s c_s$)
Z_T	Terminating impedance
	Terminating acoustic impedance ($\rho_T c_T$)
α	Absorption coefficient (nepers/m)
α_m	Maximum value of α
α_{tc}	Thermal conduction absorption coefficient
α_v	Viscous absorption coefficient
β	Phase constant
γ	Ratio of specific heats
$\mathscr{E}$	Instantaneous energy density
$\bar{\mathscr{E}}$	Mean energy density
θ	Phase angle between acoustic pressure and particle velocity
κ	Volume elasticity
λ	Wavelength
μ	Longitudinal coefficient of shear viscosity
ν	Kinematic viscosity
ξ	Particle displacement
ξ, η, ζ	Components of displacement
ρ	Instantaneous density
ρ_0	Mean value of ρ
σ	Effective scattering cross-section
τ	Relaxation time
ϕ	Velocity potential
ω	Angular frequency

CHAPTER 4

A	A function of θ_1 given by $A = \sqrt{(\sin^2\theta_1 - \eta_0{}^2)}$
a'	Volume attenuation coefficient
B	A constant given by $B = 2\eta_0{}^2 a$
b	Ratio of densities of two media
c	Propagation velocity
c_1, c_2, c_p	Propagation velocities in layers 1, 2 and p
c_v	Vertex velocity
d	Distance
	Depth of sound source
d_n	Depth of point of observation
Δd	Depth increment
g	Velocity gradient
	A function of A and B given by $(1/\sqrt{2})[\sqrt{(A^2+B^2)}-A]^{1/2}$
h	Depth of boundary between two regions of medium
	A function of A and B given by $(1/\sqrt{2})[A+\sqrt{(A^2+B^2)}]^{1/2}$
I_A, I_B	Acoustic intensities at points A and B
k	Acoustic wave number
k_1, k_2	Acoustic wave numbers in media 1 and 2
n	An integer
p	Acoustic pressure
p_i, p_r, p_t	Incident, reflected and transmitted acoustic pressures
R	Radius of curvature of sound ray
R_n	Radius of curvature of sound ray in region n
R_1, R_2	Specific acoustic resistances of media 1 and 2
s_A, s_B	Horizontal distances from sound source to points A and B
s_0, s_n, s_l	Horizontal ranges from sound source
Δs	Range increment
t	Time
u_i, u_r, u_t	Incident, reflected and transmitted particle velocities
V	Amplitude reflection coefficient
W	Amplitude transmission coefficient
x, y, z	Rectangular co-ordinates
Z_1, Z_2	Acoustic impedances in media 1 and 2
a	A constant given by $a = -a'/k$
a_r	Acoustic power reflection coefficient.
a_t	Intensity transmission coefficient
a'_t	Acoustic power transmission coefficient
a_y	Attenuation coefficient for inhomogeneous waves in Y-direction
η	Refractive index
η_0	Real part of η
θ	Angle of incidence of sound ray to normal to reflecting surface
θ_1, θ_2	Angle of incidence and refraction in media 1 and 2
λ	Wavelength
ρ_0	Density
ρ_1, ρ_2	Densities in media 1 and 2

ϕ	Phase shift upon reflection
ϕ_1, ϕ_2, ϕ_p	Angle between sound ray and horizontal in layers 1, 2 and p
ω	Angular frequency

CHAPTER 5

A	Area
a	Acceleration
B	Susceptance
B_e	Alternating component of magnetic flux density
B_0	Polarising component of magnetic flux density
C	Capacitance
C_D	Dynamic compliance
C_m	Static compliance
C_0	Static capacitance
c	Propagation velocity in air
c_b, c_e, c_ω, c'	Propagation velocity in transducer backing and driving elements, water and transducer material respectively
D	Electric flux density
d	Piezoelectric strain coefficient
d_1, d_2	Diameters of circle diagrams in air and water respectively
E	Electric field intensity
E_e	Alternating component of electric quantity
E_0	Polarizing component of electric quantity
$E(t)$	Time varying electric quantity
E_x, E_y, E_z	Components of E
e	Piezoelectric stress coefficient
F	Peak value of $F(t)$
$F(t)$	Externally applied time-varying force
f_0	Natural frequency of oscillation
f_r	Resonant frequency
G	Conductance
H	Magnetic field intensity
H_e	Alternating component of magnetic field intensity
I	Motional current
i	Current
K	A material constant
k	A constant
k_c	Electromechanical coupling factor
L	Inductance
l	Length
l, l'	Thickness of quarter- and half-wave transducer elements
m	Mass
N	Number of turns
P	Polarizing charge
P_x, P_y, P_z	Components of P
Q	Quality factor

Q_e	Electrical Q
Q_m	Mechanical Q
R	Electrical resistance
R_C	Coil resistance
R_D	Dielectric loss resistance
R_L	Mechanical loss resistance
R_m	Mechanical resistance
R_R	Radiation resistance
R_x	A parallel combination of R_L and R_R
S	Strain
s	Elastic constant
T	Stress
T_1, T_2, T_3	Tensile or compressive components of T
T_4, T_5, T_6	Shear components of T
t	Time
u	Surface particle velocity
V	Driving or applied voltage
W_e	Electrical stored energy
W_m	Mechanical stored energy
x	Displacement
x_0	Initial displacement
Y	Admittance
Z	Electrical impedance
Z_b	Characteristic acoustic impedance ($\rho_b c_b$) of transducer backing material
Z_m	Characteristic acoustic impedance ($\rho_m c_m$) of propagating medium
	Mechanical impedance
a	Transformation ratio
a_{ih}	General form of a with double subscript
β	Magnetostrictive strain coefficient
ε	Permittivity
η	Efficiency
η_1, η_2	Components of η
θ	Phase angle
κ	Volume elasticity
Λ	Magnetostrictive stress coefficient
λ	Wavelength
λ'	Wavelength in transducer material
μ	Permeability
ρ	Density
ρ, ρ_ω	Densities of air, water, transducer
ρ_b, ρ_e	backing and driving element respectively
ω	Angular frequency
ω_a	Antiresonant angular frequency
ω_0	Natural angular frequency of oscillation
ω_r	Resonant angular frequency

CHAPTER 6

$A_0, A_1 \ldots A_s$	Complex sensitivities of array elements numbered 0 to s
A_1, A_{-1}, A_2, A_{-2}	Amplitude adjustments for pattern synthesis
a	Length of each individual finite element in a linear array
D.F.	Directivity factor
D.I.	Directivity index
$D(K)$	Directional function expressed as a function of K
$D(\theta)$	Directional function expressed as a function of θ
$D(0)$	Directional function when $\theta = 0$
d	Spacing of elements of a linear point array
h	Half-width of circular transducer at distance r from centre
K	A function of θ given by $K = (2\pi/\lambda)\sin\theta$
l	Length of array
l_1, l_2	Length and breadth of rectangular transducer
l_c	Diagonal length of rectangular transducer
l_f	Diagonal length of rectangular transducer over which sensitivity is constant
m	An integer
N.F.	Noise figure
n	Number of elements in a linear point array
n'	Number of component $(\sin x)/x$ patterns used for directional pattern synthesis
r	Distance of any point on axis of array from centre
	Distance of any point on diametral axis of a circular transducer from centre
s	A number identifying each element of a linear point array
$T(r)$	Receiving array sensitivity function
$T(s)$	Relative response of array element s
t	Time
x	Rectangular co-ordinate $= (l\pi/\lambda)\sin\theta = Kl/2$
γ	Beam deflection angle
θ	Angle between incident wavefront and array axis
λ	Wavelength
ϕ	Angle between diagonal and longest side of rectangular transducer
	Phase shift interposed between signals from adjacent array elements for beam deflection
ϕ_T	Total phase shift $\phi_T = n\phi$
ω_q	Angular frequency of received signal

CHAPTER 1

Applications of Underwater Acoustics

1. General Survey

The field of underwater acoustics is so wide that it will be worth while describing some of the applications before developing the acoustic theory. In this way the theory can be more readily made compact and relevant to important practical uses; without the preliminary discussion of applications it might well lead to excessive discursiveness, both in treatment and thinking.

Applications of underwater acoustics were, until comparatively recently, largely restricted to naval operations. It was during World War I (1914–18) that a system of underwater echo-ranging was developed under the name ASDIC (said to represent the initials of the Allied Submarine Devices Investigation Committee). A better term than echo-ranging, though less used, is "echo-location". The principle was that a pulse of sound was transmitted into the water, and any reflection (or "echo") from a submarine (or other underwater object) was received by a hydrophone, which is the underwater equivalent of a microphone. The received signal could be heard on headphones, and the time delay between transmission and reception used as a measure of the range of the submarine. If the sound transmission could be made directional, then an idea of the direction of the "target" could also be obtained.

If the sound pulse were directed vertically downwards, an echo would be received from the sea-bottom, so that the depth of water beneath a ship was thus easily determinable. Such a system acquired the name "echo-sounding", and gradually came into use in purely civil applications as a navigational instrument.

Since World War II (1939–45), asdic systems have come into ever-increasing civil use in various fields of activity, notably in the fishing industry both as a research instrument and an everyday tool for locating fish shoals for catching. Other applications are in whale-catching, in hydrographic surveying, in oceanography, and in civil engineering (for observing scour around piers, etc.). Ranges obtainable vary from a few metres to many kilometres, and frequencies from a few kc/s to nearly 1 Mc/s.

In the United States of America, the term asdic became displaced by the more obvious term SONAR, which is derived from "SOund Navigation And Ranging". This term lines up very suitably with the analogous electromagnetic echo-location system called RADAR, which was developed in Great Britain between the wars (and is thus a good deal more recent than the underwater acoustic system). It is becoming evident that the name "sonar" is rapidly displacing "asdic" throughout the world, and since "asdic" has rather narrowly naval implications, it is now far preferable to use "sonar", which is a more general term, for all civil applications.

Sonar systems occur in nature. The porpoise has a very effective sonar system by which it navigates itself.[1] In air, many species of bat use sonar in place of vision, and some show a degree of refinement and accuracy which is scarcely credible and not yet fully understood.[2, 3]

Another application of underwater acoustics which has become more and more important in connection with prospecting for oil as well as other geophysical research is the use of high-power pressure pulses (e.g. generated by a small explosion) for exploring the sediments and rocks underneath the sea.[4, 5] In the simplest conception of such a system, we may think of a very high-power echo-sounder; the acoustic waves reflect, of course, from the sea-bottom, but owing to their high power penetrate the sediments and give reflections also from any discontinuities in these, e.g. from the boundary between a sediment and a hard rock. In a more complicated system the receiver may be remote from the source, so that waves detected by the former may travel by different paths, e.g. direct through the water, or partly through the water and partly through the sea-bottom. Since the velocity of propagation is different in different media (being much higher in solids than in water, and being higher in some rocks than in others), it is possible to sort out the various components of the received signal and thus learn a good deal about the nature of the sediments and rocks below the sea.

Underwater communication is possible by acoustic waves a good deal more efficiently than in air. For various technical reasons which will emerge later (mainly due to the difficulty of making efficient transducers of other than narrow fractional bandwidth) it is usual for underwater communication systems to use a modulated carrier wave. Successful systems have been developed for both speech[6] and telemetry[7] (i.e. instrument data transmission), particularly over relatively short ranges, say a few hundred metres. Speech systems are important, for example, in communicating with divers; data systems are important, for example, in oceanographical research, where instruments may be placed in midwater or on the ocean floor.

All the applications of underwater acoustics mentioned above are "active" systems, in that a source of acoustic energy forms part of the system. There are, however, many uses for a "passive" system, which involves only "listening" to acoustic emissions from independent sources. These sources may include various kinds of marine organism (some species of fish, particularly tropical ones, emit strong and characteristic noises) or man-made noises, such as cavitating propellors.

It will be clear that the term "acoustic" in this book must be taken to include the whole range of frequencies from zero up to many Mc/s, and is thus more general than the term "ultrasonic" which seems to exclude the frequency range below say 16 kc/s.

It must be appreciated that all underwater acoustic systems nowadays involve a good deal of electronic equipment, and sometimes this can be extremely complex. Some appreciation of the types of electronic system which are involved, and of the engineering concepts (such as simple information theory) which are inherent in the use of acoustic waves, is essential for the intelligent study of underwater acoustics. Thus these topics will be included in the introductory chapters of this book.

It is perhaps necessary to point out at this stage that the importance of acoustic waves for underwater detection and communication systems is mainly due to the fact that electromagnetic waves (which are more useful than acoustic waves in air over any

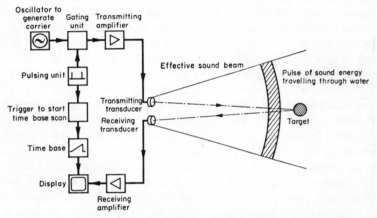

FIG. 1.1. Schematic arrangement of a typical pulsed sonar system.

distance in excess of, say, 100 m) hardly propagate at all in water except when their wavelength is so great (e.g. several kilometres) that they are useless for forming directional beams and resolving the usual objects, which have dimensions of the order of a few metres or even centimetres.

2. Sonar Systems

The majority of sonar systems, like most radar systems, use the transmission and reflection of a pulse of energy as their basis. The system is typified by the arrangement shown in Fig. 1.1.

Individual systems show many variations, such as the use of a single electroacoustic transducer for both transmission and reception, or the use of an oscillatory discharge from a capacitor into the transmitter so that no separate continuous oscillator and gating unit is required. The pulses are always bursts of carrier frequency, although the extension of the term sonar to include marine seismic work using explosive pulses is becoming more common.

The distance from the transducers to a particular reflecting object (or "target") is indicated by the time elapsing between transmission of the pulse and reception of the echo. The display therefore always includes a time base, the traverse of which is initiated by the transmitted pulse. For simple systems of the type shown in Fig. 1.1 the commonest type of display is the chemical recorder,[8] a remarkably efficient instrument which records the received information on sensitized paper, which may be a dry type such as Teledeltos, or a damp type in which the paper is impregnated with potassium iodide. In this instrument the time base is mechanical; the recording stylus is drawn across the paper. Since the mark is made when an echo is received, the position of the mark across the traverse indicates the range of the target. The paper is moved slowly in a direction perpendicular to the traverse of the stylus, so that successive traverses lie side by side. If the range of the target from the transducers does not vary, the line is produced parallel to the direction of motion of the paper. If the range changes, e.g. due to the motion of the target, or of the ship on which the equipment is fitted, then the line is sloped relative to the paper motion.

The kind of record which is obtained on a chemical recorder is illustrated in Plate I, where two portions of record taken with a vertical-beam echo-sounder are shown. The solid part represents the echoes from the sea-bottom, and as the ship has steamed along, the edge of the dark marking portrays the profile of the bottom along the ship's track. (The reason why the bottom echoes extend so far down the depth scale is simply because the sound beam was very wide, and echoes received at angles other than the vertical obviously travel a greater distance and mark at a later instant on the stylus traverse). The small dark patches above the sea-bottom represent the echoes from cod, a large species of fish of length approaching 1 m. The depth of water was actually about 220 m, but the zero of the recorder has been offset so that the chart covers a range of depth of only 100 m.

The upper record was taken in daylight, and it can be seen that the fish form agglomerated shoals. The lower record was taken at night in the same locality, and the fish are clearly seen to be spread out in a diffuse shoal.

It will be observed that the trace followed by the stylus is a circular arc. It is obviously very convenient to carry the stylus at the end of a rotating arm, and this is the most common arrangement. But there are also various kinds of straight-line recorder.

A quite different kind of display can be obtained by using the P.P.I. (Plan Position Indicator) technique, which is the normal thing in radar, but is not very common in sonar. This uses a cathode-ray tube, and the trace corresponding to a pulse travelling out along a beam in a particular direction is displayed radially from the centre of the tube, with the direction of the trace corresponding exactly with the direction of the axis of the beam relative to some datum axis, which may be either a compass bearing or the fore-and-aft axis of the ship on which the equipment is fitted. As the transducers are rotated to search in different directions, a true plan of targets is displayed. An illustration of this is shown in Plate II, where the equipment used was a fairly high-frequency, short-range type (175 kc/s, maximum range approximately 120 m). The reason this kind of display is not much used in sonar is that the velocity of sound in water is too low to permit an all-round search to be made in a reasonably short time. It is clearly necessary to wait until a transmitted pulse has travelled out to the maximum range and any echo pulse has returned before rotating the transducer to a new position. With a beamwidth of 5° and a maximum range of 750 m, say, it would require a minimum of 72 sec to sweep all round.

Another kind of sonar system (as distinct from merely a different display) which appears to correspond to that used by certain species of bat, is the "frequency-modulated" system[9] which is typified by the block schematic of Fig. 1.2. Fig. 1.3 shows the way in which the frequency changes with time. The term frequency modulation is not used in quite the same way as in a frequency-modulation communication system because the variation of frequency is imposed on the transmitter by a saw-tooth

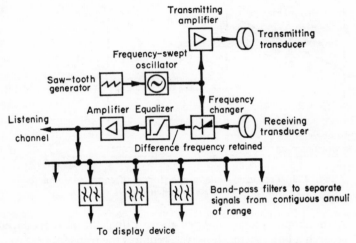

FIG. 1.2. Schematic arrangement of a "frequency modulated" sonar system.

generator and in general conveys no information. The sweep repetition period may conveniently be made twice that required for sound to travel from the transducer to the most distant range specified and back again. If the transmission is continuous as shown in Fig. 1.3, the difference-frequency output from the receiving frequency changer produced by the reflection from a fixed target is a steady frequency (e.g. f_1 or f_2 in Fig. 1.3) from the time a signal is returned from the beginning of the new sweep until the end of the sweep at the transmitter. The actual frequency of the output is clearly a measure of the range of the target, and can either be detected by the ear, or measured using a bandpass filter system. There is an ambiguity, as indicated in Fig. 1.3, at the end of each transmitter sweep, but this can be reduced in importance by making the duration of sweep long compared with the target echo time, or by making the transmission semi-continuous.

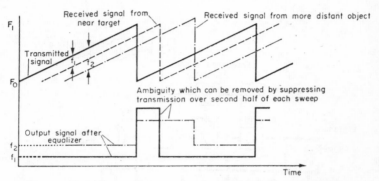

FIG. 1.3. Frequency–time graph for a "frequency-modulated" sonar system.

In the simple systems outlined above a single beam is produced. To search a large sector, however, requires the transducer(s) to be swung mechanically so that successive transmissions, or groups of transmissions "look" in different directions. As already pointed out in connection with P.P.I.-type displays, the rate of search is very slow in comparison with the corresponding rates of radar, since the velocity of sound in water is only about 1·5 km/sec.

It is increasingly being found necessary to raise the rate of scan, and electronic sector-scanning systems [10, 11, 12] have been developed with this object. In such a system the whole of the sector to be examined is insonified by a transmitted pulse which comes from a wide-beam transducer. A relatively narrow receiving beam is swung by electronic means across a sector within the time required for the pulse to travel its own length in the water. All directions within the sector are therefore effectively sampled on the one transmission. The block schematic diagram of the arrangement of one particular electronic scanning system is shown in Fig. 1.4. The receiving transducer is n times the length of the transmitting transducer, and is divided into n sections, where n is the number of beamwidths (as measured between points where the power response has fallen to half) it is desired to contain in the

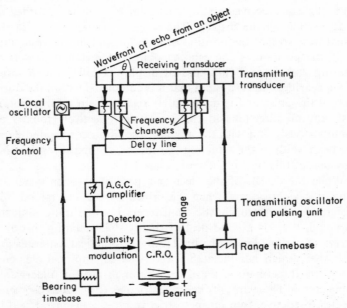

FIG. 1.4. Schematic of electronic sector-scanning sonar system.

scanned sector. If those n sections were connected to n corresponding, uniformly spaced taps on the delay line, then the beam would be deflected by an amount dependent on the phase-shift in the delay line. In the scanning system, frequency-changer equipment is inserted between the transducer sections and the delay line.

The local oscillator which feeds all the frequency-changers is swept in frequency by the bearing time-base, so that the signal frequency received by the delay line varies over a range during every sweep of the bearing time-base. If then the delay line is made to have a phase-shift which varies over this frequency range from negative values to positive values, the beam will be swept from left to right during each sweep of the bearing time-base. The latter also deflects the spot on the cathode-ray tube from left to right, so that signals received on any particular bearing are recorded on that bearing on the display. The range time-base works in the usual way, so that the position of an echo-spot on the tube indicates the position of the echoing object on rectangular axes of bearing and range. (This kind of display is called the B-scan.) If the bearing scan is so rapid that it is completed within the duration of the pulse and is immediately repeated, no information is lost, and all directions in the sector are effectively looked at simultaneously, but with the angular resolution corresponding to the beamwidth of the receiving transducer. A typical result using a system of this type is shown in Plate III.

While the design of the electronic part of a sonar system is usually straightforward when the principles are understood, the design of the electroacoustic transducers is often quite difficult. Although there is a good deal of theory[13, 14] relating to transducers of various kinds, the limitations of practical materials are such that design has inevitably a very large empirical element. There are three main types of transducer in use for underwater applications. These are: (a) magnetostrictive, (b) piezoelectric and (c) electrostrictive. In a magnetostrictive type, a magnetic field is applied to a piece of suitable magnetic material, causing the dimension of the piece to decrease along the axis parallel to the field. Thus, when the field is alternating it is necessary to polarize the material if the acoustic wave is to have the same frequency as the electric signal. The magnetostrictive effect also operates in reverse; received acoustic signals cause compression of the material which alters the magnetic field which in turn produces an e.m.f. in the electrical winding. The latter is generally a fairly thick wire with tough insulation so that the whole transducer can be directly immersed in the water. Such transducers are satisfactory for frequencies up to about 100 kc/s, and are easily constructed, although expensive.

Piezoelectric transducers use crystals in which the dimensions change according to the applied electric field. If the field is alternating the crystals vibrate and give an acoustic radiation; conversely, if the crystals are acted on by acoustic waves then they generate an electric field. Typical piezoelectric materials used for this purpose are quartz, ammonium dihydrogen phosphate, tourmaline and lithium sulphate. Whereas magnetostrictive transducers typically have low impedances of only a few ohms, piezoelectric transducers have high impedances typically of the order of 10,000 ohm and this requires high voltages for only moderate acoustic powers. They also have to be fitted in a watertight container.

Electrostrictive transducers are becoming much more widely used and seem likely to displace the other types for the majority of applications. Examples of electrostrictive materials used for transducers are barium titanate and lead zirconate in ceramic form. The change of dimensions is dependent on the magnitude but not on the polarity of the applied electric field, and thus polarization is needed as with magnetostrictive transducers. Electrostrictive transducers generally have very convenient impedances of a few hundred ohms.

With all these types of transducer it is possible to achieve efficiencies of up to 50% and Q-factors as low as about 5. (The importance of the Q-factor will become apparent in Chapter 2, since it controls the bandwidth of the system.)

The beamwidth of a transducer is inversely proportional to the dimensions of the transducer (measured in wavelengths) on the appropriate axis. Thus at low frequencies severe directional requirements necessitate very large transducers—or, more accurately, a large array of transducer elements. For this reason low-frequency echo-sounders, which normally operate at about 15 kc/s and obtain ranges of several thousand metres, operate with rather wide conical beams of perhaps 30°. High-resolution sonars, such as those used for studying the movement and behaviour of fish, need transducer arrays of tens or even hundreds of wavelengths, and therefore operate on much higher frequencies such as 400 kc/s, with ranges of perhaps 100 m. The detailed design of arrays to have particular kinds of directional patterns is a rather specialized subject, but is dealt with in its more elementary aspects in Chapter 6.

Although the design of a sonar system, as described above, involves many difficult problems and is the subject of much research, yet the greatest problems in the use of sonar undoubtedly arise in connection with the propagation of acoustic signals in the sea. (For sonar systems operating in air, corresponding problems exist but are much less well understood.)

If the medium, i.e. the sea, were loss-free, infinite in all directions and uniform in all respects, then the spreading of the sound would be spherical; in other words, the beam would expand uniformly in all directions perpendicular to its axis. Under these conditions the well known inverse-square law relates the sound intensity to the distance from the transmitter. (N.B. Sound intensity corresponds to power per unit area of cross-section). This relationship is usually expressed as a loss of 6 dB per doubling of range for one-way transmission, or 12 dB per doubling of range for the echo-signal.

The sea is not, of course, infinite in all directions, but is bounded by the surface and the sea-bed. Thus, in practice, the spreading does not follow the inverse-square law, but the sound intensity, being to some extent canalized, falls off less rapidly than the inverse-square law. In the extreme case, called cylindrical spreading, the sound intensity is proportional to the reciprocal of the range, i.e. the loss is 3 dB per doubling of the range for one-way transmission.

In addition to this loss due to spreading, there is an additional loss due to absorption (and conversion into heat) of the sound energy by the water. This loss is, as would be expected, very variable. It is small at low frequencies but rises very rapidly with increase of frequency, and over the range of frequencies in common use at present, the loss expressed as dB/km is approximately proportional to the square of the frequency. At 50 kc/s the absorption loss varies from about 8 to about 16 dB/km.

Further propagation effects which are of great importance are caused by variations of the velocity of sound from one part of the sea to another. Under normal conditions the velocity of sound in the sea is about 1500 m/sec. The main causes of variation of this velocity are, in the usual order of importance: temperature, pressure due to depth, and salinity. The actual magnitude of the effects can be indicated by the convenient empirical formula of Wood[15] which applies in the temperature range 6–17°C:

$$c = 1410 + 4 \cdot 21t - 0 \cdot 037t^2 + 1 \cdot 14s + 0 \cdot 018d,$$

where c = velocity in m/sec, t = temperature in degrees C, s = salinity in parts per thousand, d = depth in m.

Perhaps the most serious effect of variations of velocity is the refraction of the sound beam when velocity changes with depth. This could cause a beam, perhaps normally horizontal, to be deflected to the sea-bottom, where it will be reflected upwards only to be refracted down again, and so on. Along any straight line through the transducer, therefore, there are intervals of range where detection is impossible and others where it is possible. This kind of effect is, of course, more serious with low-frequency systems as they have the greatest nominal range.

Another effect of variations of velocity is the general scattering of the sound beam as it passes through turbulent regions; this leads to rapid fluctuation of received signal strength.

When the object of the sonar system is to detect small targets (as distinct, for example, from the sea-bottom) the target signal has to be detected against a random background due to "reverberation"; this is the sum of all the numerous small echoes produced by backscattering from sand and stone particles on the sea-bottom, from minute air bubbles and other inhomogeneities in the water, from waves on the surface and so on. Although background noise in the sea, for example, noise due to waves breaking, etc., may limit the maximum range of detection, reverberation, on the other hand, may be a limitation of performance at all ranges. This is because it arises from the signal transmission and at any time interval thereafter has a power level closely related to that of the signal. Thus for detection of small objects in areas where reverberation level is high, due, for example, to a shallow rough sea-bottom, it is essential to use beams which are as narrow as possible in the relevant dimension. Because reverberation is normally the limiting factor in detection, an increase of transmitted power is not effective in improving detection over most of the range; consequently most sonar systems operate at acoustic powers of fairly low peak level, e.g. 1 W–1 kW.

3. Seismic Systems

In the sonar systems described in the previous section, the signals have been confined to a frequency band centred around a carrier frequency, and the bandwidth has been a small fraction of the centre frequency. Even in an ordinary sonar system there are advantages in increasing this fractional bandwidth in order to improve target resolution while still retaining a fairly low carrier frequency with its relatively low propagation losses. Moreover, specially wide band sonar systems (e.g. ratio of upper to lower frequencies of perhaps 10) may have advantages in permitting the display of frequency-response of targets; for example, it is thought possible that this would help the identification of fish echoes. But when we come to systems (reasonably called seismic systems) intended to explore the sediments and rocks beneath the sea, it is necessary to use pulses of a different type. These are basically unidirectional (or "d.c.") pressure pulses generated by a single short impact, such as may be produced by detonating an explosive charge, by producing an underwater arc, or by a powerful "thump" perhaps generated electromagnetically. Such pulses contain energy at frequencies from zero up to quite high frequencies.

The advantages of these explosive-type pulses for geophysical purposes are, firstly, that they can be generated cheaply at very high energy levels (for which electronic generation would be out of the question), and secondly, that they contain dominantly the very low frequencies which propagate with minimal losses. There are no known electroacoustic transducers suitable for the transmission of this kind of frequency range and energy level with reasonable efficiency.

On reception, of course, either delicate mechanical instruments of the geophone type, or hydrophones followed by electronic amplifiers, have to be used. In the latter case, the frequency range

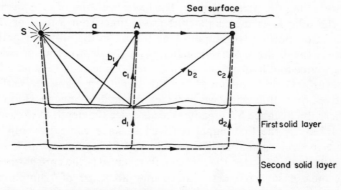

FIG. 1.5. Showing sound paths between explosive sources and hydrophones A and B.

involved (being both very low, and of large ratio of upper to lower frequencies) is a very great difficulty and involves the use of aperiodic transducers with large damping losses. The higher attenuation of the higher frequencies in the several media cause the received pulse to have a much slower rise than that transmitted, and it is often difficult to distinguish the various returns over the different paths.

Provided the received pulses can be correctly identified, the way in which the system works may be illustrated as follows. In Fig. 1.5 is shown a simple representation of a vertical section through the sea and its underlying rocks, and here S represents a sound source, and A and B are two different positions of a receiving hydrophone. The simplest path by which the sound from S can reach A and B is the direct path a. This is purely in the water. Another path is b_1 to A or b_2 to B. This, although entirely in the water, involves a reflection at the sea-bed, and is always of greater travel time than a. Sound from S can also enter the first solid layer (usually a sediment) and be refracted, due to the greater velocity of propagation in the solid, along a horizontal path. Sound escapes from this path all the way along, and any reaching A or B has to suffer a further refraction on entering the water along the paths c_1 or c_2. The higher velocity of propagation in the solid layer means that the travel time from S to A or B is less in proportion to the path length than in the all-water paths. Where the distance SA is less than twice the depth of water below the source and hydrophone (which are both usually very near the sea-surface), it is obvious that the signal travelling by path a arrives first. But when we consider the hydrophone position B, where the distance SB greatly exceeds twice the depth, it is clearly possible for the first signal received to have come by path c_2. Thus, provided suitable distances SA and SB are taken, and knowing the depth of water, the first signals received (which are naturally the most easily recognized, since they are clear of "rumblings" due to secondary effects of the explosion and to miscellaneous scattering paths) enable the velocity in the solid to be calculated to a quite reasonable accuracy. This gives considerable information about the nature of the rock. Other layers of rock underlying the first are normally harder and have higher sound velocities, and so also give refracted paths as marked d_1 and d_2 in Fig. 1.5. Over even greater distances, therefore, these paths can give the first received signal, and so their velocity too can be measured. It is thus possible to build up a picture of the geological features below the sea, different layers being identified over a large region by their sound velocity. Certain kinds of profile are recognized by geophysicists as having special significance, e.g. as being likely to contain an oil-field.

The velocity of propagation of sound, which is about 1.5 km/s in water, may range from only a little above this figure for some

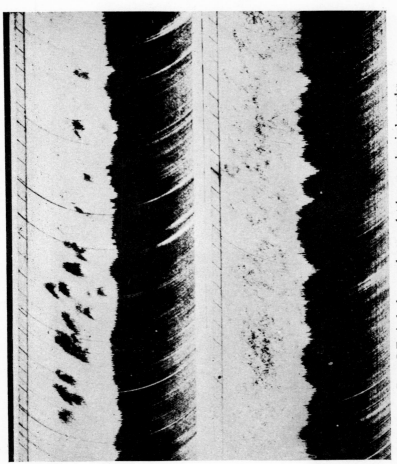

PLATE I. Typical echo-sounder records taken on a chemical recorder.

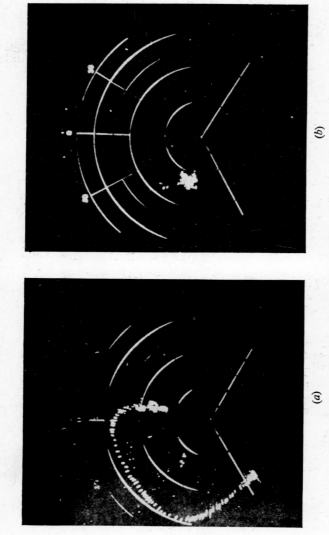

PLATE II. Two P.P.I. displays obtained on the Sea-scanner equipment: (a) a purse seine net being towed, with a few salmon in it, (b) a small shoal of herring. The pattern of circles and lines is that of the range and bearing markers.

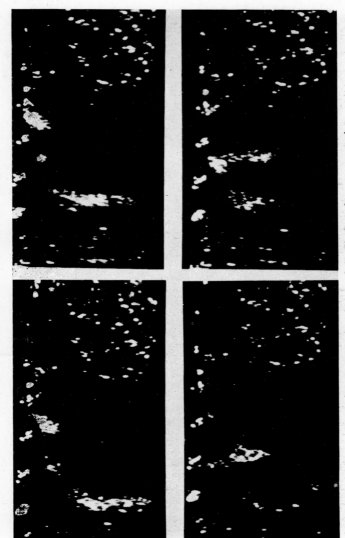

PLATE III. Typical photograph of the display of a sector-scanning sonar. (Taken at 5 sec intervals these show motion of small fish shoals relative to a line of pier supports.)

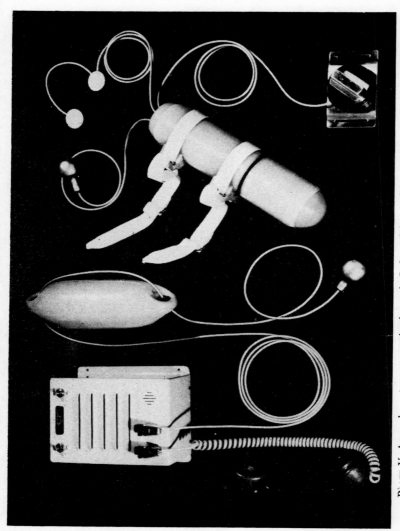

PLATE V. An underwater communication unit. Left to right: surface unit, ultrasonic transducer, bone-conduction head-set and on–off transmit–receive switch.

clays up to well over 6 km/s for some granites and other very hard rocks.

The sort of record which is obtained in this kind of work is shown in Plate IV; this is, however, for a particularly simple situation in which only one path through the solid sea-bottom is observable. The first arrival occurs about 2·5 sec after the explosion; this corresponds to path c_1 in Fig. 1.5. The direct wave via path a arrives about 1·6 sec later.

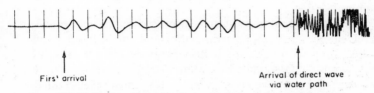

First arrival

Arrival of direct wave
via water path

PLATE IV. Typical marine seismic record for a simple situation with a single rock path.

4. Underwater Communication and Telemetry Systems

Apart from the fact that the main transmission link is acoustic, these systems are basically the same as those using electromagnetic signals either on lines or radio, i.e. they involve the transmission of information as a modulation of a carrier tone or pulse sequence. It will be sufficient for the purposes of this book to describe a simple speech communication system recently developed for guiding divers,[6] a depth-of-net telemetry system,[7] and the nature of the propagation difficulties.

In the speech system, the diver uses a full-face breathing mask to enable him to speak normally underwater and a throat microphone to obtain his speech signals. These are used to modulate the frequency of the transmitted acoustic carrier tone which has a centre frequency of 120 kc/s. This frequency-modulated tone is radiated at an acoustic power of 100 mW using a tubular electrostrictive ceramic transducer which is omnidirectional in the horizontal plane. Its vertical 3 dB beamwidth is restricted to 30°, however, to prevent signals reaching the surface of the sea; they might otherwise be reflected from suitably orientated wave surfaces and arrive at the receiving transducer in addition to the signal travelling along the direct path. Each of these surface-reflected signals would have travelled via a path of different length and would be shifted in phase relative to each other and to the direct signal. As a result of this multipath propagation, distortion of the speech signal would occur. A restricted vertical beam reduces multipath effects to a minimum and ensures reliable midwater communication.

The transmitting transducer is also used for reception and the received f.m. signal after discrimination and amplification is passed to the diver via a bone-conduction earpiece. Plate V illustrates a typical self-contained underwater unit comprising a transmitter, a receiver, and also a voice-operated switch which relieves the diver of the task of manually switching from the transmit to the receive positions. Such a system permits multiway underwater communication up to ranges of at least 1 km at depths in excess of 70 m.

In underwater communication systems frequency modulation is employed in preference to amplitude modulation owing to its better discrimination against noise. For this reason it is also used in systems designed to transmit data from submerged instruments over what is called a "telemetry" link. One such system, designed to indicate automatically the depth of a fishing net, employs a transmitting unit containing a temperature-compensated parallel-plate capacitor, the capacitance of which is dependent upon external pressure and hence its depth. This capacitor forms part of the tuned circuit of an oscillator, and thus the frequency of oscillation is dependent upon pressure. The transmitting unit which is attached to the trawl warp near the mouth of the net also contains an amplifier and a transducer and relays its depth information as an acoustic signal whose frequency is proportional to depth. It is received by another transducer attached to the trawl warp just below the sea-surface and fed to a frequency discriminator and depth indicator unit on the bridge of the trawler. Both transmitting and receiving transducers have relatively narrow beamwidths to discriminate against noise generated by surface waves and the trawler itself, but they are automatically aligned by being attached to the trawl warp as illustrated by Fig. 1.6.

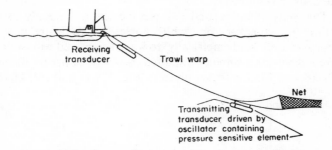

FIG. 1.6. Showing arrangement of a depth-of-net telemeter.

5. Passive or "Listening" Systems

Most of the noises (as distinct from deliberate signals) which it is required to observe underwater are of wide frequency spectrum. Thus the main problem is to find hydrophones of reasonable sensitivity over a wide frequency band. This is very similar to the hydrophone problem in seismic systems, and is solved in the same sort of way.

For the noises made by marine organisms,[16] the frequency range of interest extends from a few c/s up to tens of kc/s. Noise from other sources, natural or man-made, similarly extends over a very wide spectrum; but it is almost universally true that in the upper frequency ranges, the sound intensity at any particular point decreases with frequency, roughly according to an inverse-square law. This presumably is closely connected with the fact that in the same frequency ranges the absorption loss (measured in decibels) of sound energy in the water increases approximately as the square of frequency.

In measuring noise levels, it is usual practice to quote results as sound pressure level in a specified frequency band, in decibels relative to a datum of 0·0002 dynes/cm². In order to normalize the results, it is usual to quote the spectrum density as pressure level in a band of 1 c/s width centred on the frequency for which the density is quoted.

The general background noise in the sea, due mainly to wave action (e.g. breakers or "white horses"), can be related very approximately (and empirically) to wave height and wind velocity. Some typical measurements quoted in a basic paper by Knudsen, Alford and Emling[17] are given in Fig. 1.7(a) and (b). Here the

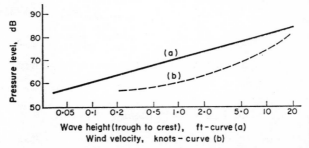

Fig. 1.7. Overall sound pressure level in band 0·1–10 kc/s as a function of (a) wave height, (b) wind velocity.

noise levels are the overall levels measured in a frequency band of 0·1–10 kc/s. A typical spectral response of this kind of noise is shown in Fig. 1.8 curve (a) from which it will be seen that the spectral pressure level falls off at about 5 dB per octave. It will be shown in Chapter 3 that sound intensity or power is proportional to the square of the sound pressure; therefore the falling-off of sound intensity is inversely proportional almost to the square of the frequency. Since this kind of noise is so closely related to the state of the sea, it is frequently called "sea-state noise".

Noise from marine animals is very variable.[17, 18] The spectral and temporal characteristics may be distinct enough to permit identification of the species in some cases. Animals of temperate and colder waters seem to make less noise than those of tropical waters. Two typical spectra are shown in Fig. 1.8. Curve (b) relates to the snapping shrimp, which occurs in waters with minimum temperatures above about 12°C and at depths not greater than about 60 m. The spectral level does not start to fall until above about 10 kc/s. Curve (c) relates to the croaker, which occurs in great concentration on the eastern coasts of the U.S.A. Here the spectral level falls off above about 0·5 kc/s.

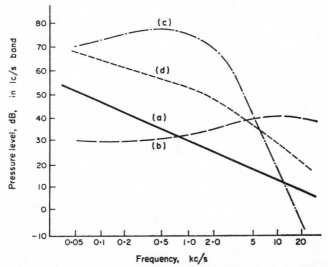

Fig. 1.8. (a) Spectral response of sea-state noise under fairly calm conditions. (Overall pressure level in band 0·1–10 kc/s was 68 dB.) (b) Spectral response of noise from snapping shrimps. (c) Spectral response of noise from croakers. (d) Spectral response of noise in a busy harbour entrance. (Overall pressure level in band 0·1–10 kc/s was 90 dB.)

Noise produced by man-made sources, e.g. ship's machinery and propellor cavitation, is naturally even more variable. But it does seem characteristic that the very low frequencies (e.g. 0·1 kc/s) are dominant. A typical spectrum given in the paper cited above is shown in curve (d) of Fig. 1.8; this relates to the entrance to a busy harbour.

The detection and location of localized sources of noise (e.g. a ship or a fish shoal) require special techniques. Ordinary hydrophones are not made markedly directional. It will be seen that, as the beamwidth of a transducer is normally inversely proportional to frequency, it is not usually satisfactory to make hydrophones large enough to be directional, since directional properties dependent on frequency lead only to confusion when very wideband noise is to be located and measured. Thus ordinary hydrophones are not suitable for location of a noise source. It is possible, by using specially shaped surfaces in the hydrophones, or by using complex electrical circuits with a multi-element hydrophone, to achieve a constant beamwidth over a wide frequency band.[19, 20, 21] It is also possible to use correlation techniques;[22] the concept of "correlation function", which is involved in these, is described in Chapter 2.

References

1. KELLOGG, W. N., *Porpoises and Sonar*, Phoenix Science Series, University of Chicago Press (1962).
2. GRIFFIN, D. R., *Listening in the Dark*, Yale University Press (1958).
3. KAY, L., Orientation of Bats and Men by Ultrasonic Echo Location, *Brit. Communications and Electronics* 8, 582 (1961).
4. SHEPARD, F. P., *The Earth Beneath the Sea*, Johns Hopkins Press, Baltimore (1959).
5. GASKELL, T. F., *Under the Deep Oceans*, Eyre and Spottiswoode, London (1960).
6. GAZEY, B. K., and MORRIS, J. C., An Underwater Acoustic Telephone for Free-swimming Divers, *Electronic Eng.* 36, 364 (1964).
7. TUCKER, M. J., The N.I.O. Depth Telemeter, *Proc Int. Telemetering Conference* 1, 224 (1963).
8. GRIFFITHS, J. W. R., and MORGAN, I. G., The Chemical Recorder and its Use in Detecting Pulse Signals in Noise, *Trans. Soc. Inst. Tech.* 8, 62 (1956).
9. KAY, L., An Experimental Comparison Between a Pulse and a Frequency-modulation Echo-ranging system, *J. Brit. Inst. Radio Engrs.* 10, 785 (1960).
10. TUCKER, D. G., WELSBY, V. G., and KENDELL, R., Electronic Sector Scanning, *J. Brit. Inst. Radio Engrs.* 8, 465 (1958).
11. TUCKER, D. G., WELSBY, V. G., KAY, L., HENDERSON, J. G., TUCKER, M. J., and STUBBS, A. R., Underwater Echo-ranging with Electronic Sector Scanning: Sea Trials on R.R.S. Discovery II., *J. Brit. Inst. Elect. Engrs.* 19, 681 (1959).

12. WELSBY, V. G., and DUNN, J. R., A High-resolution Electronic Sector-scanning Sonar, *J. Brit. Inst. Radio Engrs.* **26**, 205 (1963).
13. HUNT, F. V., *Electroacoustics: The Analysis of Transduction, and its Historical Background*, Harvard Univ. Press (1954).
14. HORTON, J. W., *Fundamentals of Sonar*, U.S. Naval Inst. Annapolis (1957).
15. WOOD, A. B., *A Textbook of Sound*, Bell, London (1949).
16. TAVOLGA, W. N. (editor), *Marine Bio-Acoustics*, Pergamon, Oxford (1964).
17. KNUDSEN, V. O., ALFORD, R. S., and EMLING, J. W., Underwater Ambient Noise, *J. Marine Res.* **7**, 410 (1948).
18. FISH, M. P., KELSEY, A. S., and MOWBRAY, W. H., Studies on The Production of Underwater Sound by North Atlantic Coastal Fishes, *J. Marine Res.* **11**, 180 (1952).
19. TUCKER, D. G., Arrays with Constant Beamwidth Over a Wide Frequency-range, *Nature* **180**, 496 (1957).
20. MORRIS, J. C., and HANDS, E., Constant-beamwidth Arrays for Wide Frequency Bands, *Acustica* **11**, 342 (1961).
21. MORRIS, J. C., Broad-band Constant-beamwidth Transducers, *J. Sound Vib.* **1**, 28 (1963).
22. NODTVEDT, H., The Correlation Function in the Analysis of Directive Wave Propagation, *Phil. Mag.* **42**, 1022 (1951).

CHAPTER 2

Basic Problems of Signal Reception, Detection and Display

1. General

The discussion of underwater acoustic systems in Chapter 1 showed the basic conceptions of the systems and mentioned some of the difficulties which arise in practice; but little was said regarding the fundamental limitations to performance and the factors determining the design parameters. The matters involved are part of what is known as information theory (or sometimes, in this kind of context, information engineering), and are fully discussed in a rigorous mathematical manner in many text books.[1, 2, 3, 4] Here we shall give a simplified account of what is necessary as a background to applied underwater acoustics.

2. Bandwidth

One of the design factors which appears fundamental in this work is the bandwidth of the system. It is shown in books on electrical network theory[5, 6] that the rate at which a circuit can respond to a suddenly-applied signal is proportional to its bandwidth expressed as the difference between the upper frequency limit and the lower. Of course, there are many complications to this simple conception; for example, practical circuits do not have a sudden change of response such that above a certain frequency signals are completely suppressed and below it are transmitted without loss or distortion. Thus a definition of effective bandwidth is needed for rigorous work, and it can be given in a number of ways according to the circumstances; but a definition adequate for present purposes is the difference between the upper and lower frequencies at which the response is 3 dB below that at the centre frequency. This is called the "3-dB bandwidth". It is illustrated in Fig. 2.1. A circuit with a response like this is called a "bandpass

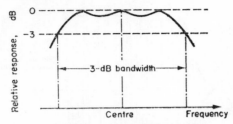

FIG. 2.1. Illustrating 3-dB bandwidth.

circuit". When the circuit transmits down to zero frequency (or d.c.) as illustrated in Fig. 2.2, the term bandwidth can be misleading, and it is better to refer to the "3 dB-cut-off frequency". A circuit with a response like this is called a "low-pass circuit". (N.B. 3 dB corresponds to a power ratio of 2 : 1).

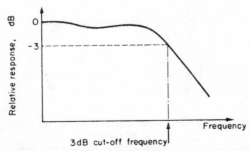

FIG. 2.2. Illustrating 3-dB cut-off frequency for circuit transmitting down to zero frequency.

Another complication is that the nature of the response to a suddenly-applied signal depends on the nature and frequency of the signal.[7] But for a suddenly-applied tone at the centre frequency of the response shown in Fig. 2.1 the output waveform would be roughly of the shape shown in Fig. 2.3; and for a d.c. suddenly applied to the low-pass circuit of Fig. 2.2, the output waveform would be roughly as shown in Fig. 2.4. The rate of

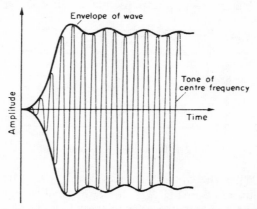

FIG. 2.3. Output waveform from band-pass circuit when a tone at the centre frequency is suddenly applied at zero time.

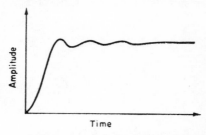

FIG. 2.4. Output from low-pass circuit when d.c. is suddenly applied at zero time.

rise of the response (taking the envelope in the bandpass case) can be measured as the slope of the curve at half the final amplitude; and if this is normalized by taking the final amplitude as unity, and if time is measured in seconds, then the rate of rise is almost always between 1 and 1·6 times the 3-dB bandwidth measured in c/s for the bandpass circuit, and between 2 and 3·2 times the 3-dB frequency (in c/s) for the low-pass circuit.

The relation of this matter to simple sonar systems lies mainly in the resolution of objects in range. If the outgoing pulse encounters a very small reflecting object at a particular range R_1, then a sudden small burst of the transmitted frequency is returned to the receiver. If this latter has a very wide bandwidth, it responds instantly, both to the sudden growth of the echo-signal, and to its sudden decay, and the returning pulse is correctly reproduced. If, in addition to the object at range R_1, there is another small object at a very slightly greater range R_2, then this too will give a correctly reproduced echo pulse, and the two objects can be separated by the receiver if $R_2 - R_1 > L$, where L is the length of the acoustic pulse in the water. Here $L = cT$, where c is the velocity of propagation of sound and T is the time-duration of the transmitted pulse.

Now suppose that the bandwidth of the receiver is not very wide, and is, in fact, restricted so that the rise-time (i.e. the reciprocal of the rate of rise previously discussed) is much greater than T. Then the pulses received are not correctly reproduced, although if there is only one echoing object it can perhaps be detected. If there are the two objects close together as before, then their echo pulses are so spread out by the receiver that they overlap and the two objects cannot be separated, or "resolved", by the receiver.

A similar effect occurs if the transmitter has too narrow a band and so emits a badly shaped and elongated pulse into the water. In general, it can be said that the sonar system must have an overall bandwidth (in c/s) at least of the order of c/L, where c is in m/sec and L (in metres) is the minimum spacing (in range) of targets which it is required to separate.

Another reason for requiring a wide bandwidth in a sonar receiver is to accommodate the shift of frequency due to the Doppler effect if the echoing objects are moving. But it is undesirable to have too wide a band, as this allows too much noise to be received and thus spoils the detection of the objects. In practice, bandwidths are generally limited by the transducers, since it is difficult to get efficient transducers with a bandwidth of more than say one-fifth of the centre frequency.

In the case of an electronic sector-scanning sonar system in which the receiving beam is caused to scan over a sector of n beamwidths within the duration of the pulse used to insonify the sector, there is an additional bandwidth requirement. At the receiving transducer itself, the situation is no different from that of a non-scanning sonar, and the transducer bandwidth has to be merely of the order of c/L as above. But after the electronic scanning process, the situation is quite different. The system has been caused, in effect, to take short samples of the pulse returns from n independent directions.[8] Although the scanning may, in fact, be continuous, it can be shown that only n separate samples of the signals in the sector are needed to define them fully. If, therefore, we regard the signals, after scanning, as crudely equivalent to a sequence of sample pulses each of duration $1/n$ of that of the transmitted pulse, we see that a bandwidth of n times that needed at the transducer has now to be provided. This result can also be obtained rigorously without invoking the idea of sampling. A useful way of looking at this result is that as the information rate has been increased by a factor of n (due to scanning), the bandwidth also has to be increased by the same factor.

The importance of bandwidth in seismic systems is similar, but the resolution required is not between the echoes from different objects; it is, of course, between the signals transmitted over different paths. The receivers are, in general, more of the low-pass type than the bandpass, but cannot respond to zero frequency. Since the transmissions cover a wide band of frequencies (due to their explosive nature), the problem on reception is to select some part of the frequency spectrum which gives the best results; it is not possible to utilize all the transmitted spectrum as the high frequencies suffer high attenuation. Generally, the receiver has a choice of cut-off frequencies available, and the choice is made empirically.

The importance of bandwidth in communication and telemetry

systems is primarily in regard to the rate at which information may be transmitted. In a speech system, of course, the primary necessity is to transmit most of the frequency spectrum actually generated by the telephone transmitter—usually about 300–3000 c/s. When the system uses a modulated carrier of some sort, the minumum theoretical bandwidth required in the medium is the 2700 c/s corresponding to the bandwidth occupied by the original signal. Using single-sideband techniques (as in line telephony, for example) this could nearly be achieved; but it is usual to use envelope-modulation or frequency-modulation (as in the example in Chapter 1), which require at least twice the original bandwidth. For fuller details of this matter a textbook on modulation systems should be referred to.[3, 9, 10]

In a telemetry system, transmitting data say in the form of pulses, the relation of bandwidth to information rate is clearly dependent on the relation between bandwidth and rate of response in the system, previously discussed. It is necessary to separate, or resolve, the individual pulses, which carry individual information; and the bandwidth must therefore clearly be great enough to permit each pulse to rise and decay substantially before the next is received. The minimum workable bandwidth in c/s is very approximately of the order of the pulse rate in pulses/sec.

3. Noise

Although it is often useful to discuss the way in which systems operate under conditions when the wanted signals are the only ones present—as in the previous section—yet, in practice, these conditions are never attainable. In all circumstances there are present irrelevant disturbances; some of these may be interference from other systems, others may be a function of the system itself (e.g. reverberation, mentioned briefly in Chapter 1), and others are inherent in any system. In the last class we include all effects due to the discrete nature of matter and natural phenomena, and these effects are generally called "noise". The most

basic noise is probably that due to the random motion of electrons
in a conductor; this leads to random fluctuations in the charge
density of electrons, and so to random variations in the potential
at any point in the conductor. Thus at the terminals of a conductor of resistance r ohms, and at absolute temperature T
degrees, a fluctuating voltage is obtained, the mean-square value
$(\overline{e^2})$ of which is proportional to both r and T. Johnson[11] showed
experimentally that

$$\overline{e^2} = 4kTrB \qquad (2.1)$$

or in other words the noise power N.P. is given by

$$\text{N.P.} = 4kTB \qquad (2.1a)$$

where k is a constant called Boltzmann's constant, and B is the
effective bandwidth of the system in c/s. (This effective bandwidth is not, in general, quite the same as the 3-dB bandwidth
discussed in Section 2, but is usually near enough to it for most
practical purposes.) This kind of noise is usually called Johnson
or resistance or thermal noise. Although the magnitude depends
on bandwidth, it does not depend on the centre frequency.

Other sources of noise arise in electronic devices, such as shot
noise[11] in thermionic valves. These do not all have the kind of
relationship shown by eqn. (2.1). For example, noise in tran-
sistors depends not only on the bandwidth of the circuit, but also
on the actual frequencies involved.

In an underwater acoustic system, noise of the thermal type
arises in the water itself. This shows in the receiver as a voltage
across the transducer terminals given exactly by eqn. (2.1), where
r is taken as the radiation resistance of the transducer (see Chapter
5 for a full account of radiation resistance; for now, it may be
taken as the additional electrical resistance which is measured due
to the damping of the transducer motion by the water) and T
is the absolute temperature of the water (not of the transducer).

The kinds of noise mentioned above are, of course, of very
low level and become noticeable only when large amplification is
used in a receiver. But they do obviously constitute a fundamental
limitation to the sensitivity of any receiver or system. Once the

wanted signal level has sunk to the noise level, no amount of amplification can make it any more distinguishable.

In the sea there are many other kinds of noise, due to natural causes and to animal and human activity. For the most part, and especially when taken all together, these have at least one thing in common with thermal noise; namely, they are random. By random we mean that the exact amplitude of the disturbance at any instant is unpredictable. But if the noise is "stationary", the statistical parameters of the noise do not vary with time, so that the average and mean-square values are constant and known. Most kinds of noise, and certainly thermal noise, have a particular kind of relationship between the probability of the particular amplitude occurring and the amplitude itself called the Gaussian relationship;[11] such kinds of noise are often referred to as "Gaussian noise". This relationship is

$$p(x) = A \exp \left[-(x_0 - x)^2 / 2\sigma^2 \right] \qquad (2.2)$$

where $p(x)$ is the relative probability of an instantaneous amplitude x occurring, x_0 is the mean value of x, and σ is a parameter called the "standard deviation" which measures the spread of

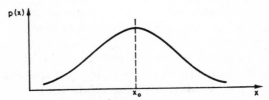

FIG. 2.5. General shape of Gaussian distribution of probability $p(x)$ of amplitude being x.

amplitudes. A is a scaling factor. The general shape of this distribution is given in Fig. 2.5. Since the total probability of all amplitudes of x must be unity, it is clear that

$$\int_{-\infty}^{\infty} p(x) \, . \, \mathrm{d}x = 1 \qquad (2.3)$$

and this defines A as

$$A = \frac{1}{\sigma\sqrt{(2\pi)}}. \tag{2.4}$$

If noise of this kind is received in a bandpass system, there is clearly no d.c. component, so that $x_0 = 0$, and σ = root-mean-square (r.m.s.) amplitude. Under these circumstances the noise waveform can be represented in a number of simple ways. When the bandwidth is a small fraction of the centre frequency, ω_0, the following representation is often useful:

$$v_N = A(t) \cos [\omega_0 t + \phi(t)] \tag{2.5}$$

where $A(t)$ is an amplitude varying randomly with time and $\phi(t)$ is a phase angle varying randomly with time. Rather more generally, in terms of bandwidth, the noise amplitude may be represented as

$$v_N = \sum_{r=1}^{r=n} a_r \cos (\omega_r t + \phi_r); \ n \to \infty \tag{2.6}$$

where ϕ_r is a random sequence of numbers, but the a_r and ω_r are spaced over the available bandwidth to give the correct spectral distribution.

In making calculations of the performance of underwater acoustic systems (such as sonar, in particular) in respect of noise interference, it is often not necessary to have a rigorous representation of noise. The object is usually to determine a signal-to-noise ratio in terms of r.m.s. values. In these circumstances, a non-rigorous representation of noise may be used as follows:

$$v_N = a \sum_{r=1}^{r=n} \cos \omega_r t; \ n \to \infty \tag{2.7}$$

where the ω_r are distributed over the bandwidth to give the required power spectrum (usually uniform). This representation has many of the characteristics of true random noise.[12] For instance, at any unspecified time the probability of obtaining any particular amplitude follows the Gaussian law of eqn. (2.2); the voltage can be represented by a r.m.s. value. Its limitations arise from the fact that at a *specified* time (especially at $t = 0$) its value is predictable; but this does not matter in the kind of calculations to which we shall be restricted in this book. We shall, therefore, use the representation of eqn. (2.7) when we make noise calculations.

The noise of eqn. (2.7) has a r.m.s. voltage V_N given by

$$V_N = a \left\{ \overline{\left[\sum_{r=1}^{r=n} \cos^2 \omega_r t \right]} \right\}^{1/2} \tag{2.8}$$

where the bar above the bracketed term means "average value". Therefore

$$V_N = (1/\sqrt{2}) \, a\sqrt{n} \tag{2.9}$$

If there is also a signal $E \cos \omega t$, which has a r.m.s. voltage of $E/\sqrt{2}$, the *r.m.s. signal-to-noise ratio*, R, is clearly

$$R = \frac{E}{\sqrt{2}} \bigg/ V_N = \frac{E}{a\sqrt{n}} \tag{2.10}$$

4. Correlation

Sonar systems, in particular, often have more than one receiving transducer element and more than one channel in the receiving electronic system. The relationship between the noise in one channel and that in another is then important, and this can be measured by the *correlation* between them. Let the noise in channel 1 be

$$v_{N_1} = a_0 \sum_{r=1}^{r=n} \cos \omega_r t + a_1 \sum_{p=1}^{p=m_1} \cos \omega_p t \tag{2.11}$$

and that in channel 2 be

$$v_{N_2} = a_0 \sum_{r=1}^{r=n} \cos \omega_r t + a_2 \sum_{q=1}^{q=m_2} \cos \omega_q t \tag{2.12}$$

where the ω_p and ω_q are quite different frequencies.

Clearly, then, the two noise voltages have a part in common and a part different. They are thus said to be partially correlated.

The actual degree of correlation is measured by the "correlation coefficient", ψ, defined as

$$\psi = \frac{\overline{v_{N_1} \cdot v_{N_2}}}{V_{N_1} \cdot V_{N_2}} \tag{2.13}$$

where the average of $v_{N_1} \cdot v_{N_2}$ is taken, in principle, over an infinite time as in taking the r.m.s. values V_{N_1} and V_{N_2} by the method of eqn. (2.8). Thus, if the noises are fully correlated, so that $a_1 = a_2 = 0$, then

$$\psi = \frac{a_0^2 \cdot \frac{1}{2}n}{a_0^2 \frac{1}{2}n} = 1 \tag{2.14}$$

so that unity correlation factor indicates full correlation. If the noises are quite uncorrelated, so that $a_0 = 0$, then

$$\psi = \frac{a_1 a_2 \left[\sum\limits_{p=1}^{p=m_1} \cos \omega_p t \cdot \sum\limits_{q=1}^{q=m_2} \cos \omega_q t \right]}{a_1 \sqrt{(\frac{1}{2}m_1)} \cdot a_2 \sqrt{(\frac{1}{2}m_2)}}$$
$$= 0 \tag{2.15}$$

since all the ω_p and ω_q are different. Thus, as would be expected, the correlation coefficient is zero if there is no correlation.

For partial correlation, clearly ψ lies between 0 and 1.

The importance of this matter can be seen by a simple example. Suppose we have s identical transducer elements and channels, all with the same received signal $E \cos \omega t$, and with the same r.m.s. noise voltage V_N. The signal-to-noise ratio on each channel is $E/\sqrt{(2)} V_N = R_1$, say. Now the channels are later combined by adding together their voltages. If the noise waveforms are all correlated, then the total r.m.s. noise voltage is sV_N and the total signal voltage is sE. The output signal-to-noise ratio is therefore R_1, just as on the individual channels. But if the noise waveforms are all uncorrelated, then the total r.m.s. noise voltage is $\sqrt{(s)} V_N$ although the total signal voltage is still sE. The output signal-to-noise ratio is now $\sqrt{(s)} . R_1$, an improvement of $\sqrt{s}$.

Whether noise in different channels is fully correlated, partially correlated, or uncorrelated depends on a variety of factors. If it originates as thermal noise in different resistance elements, there will be no correlation. If it originates as, say, "sea-state noise" (i.e. noise due to waves breaking in the sea-surface and being well spread in direction of arrival) and is received by transducer elements spaced apart by several wavelengths, then again there will be substantially zero correlation; but if the transducer elements are very close together (say less than half-a-wavelength apart) there will be substantially full correlation. If the noise originates in one particular direction (and is thus directional in the same way as the signal), there may be full correlation, and almost certainly partial correlation.

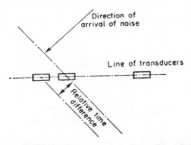

FIG. 2.6. Noise wave arriving at line of transducers.

This last example conveniently introduces the idea of *correlation function*. Suppose the transducer elements lie in a straight line, as shown in Fig. 2.6. If the source of noise in the sea lies in a particular direction which is normal to the line of transducers, and is far enough away for the wave to be plane, then clearly all transducer elements receive an identical waveform and there is full correlation. But if the direction is not normal, then each element receives a waveform which is the same in shape but shifted in its time scale relative to the others. This reduces the correlation, as may be seen by considering just one particular frequency component, say $a_r \cos \omega_r t$. Between one element and its neighbour there will be a phase difference ϕ_r due to the extra travel time. So we have

$$a_r \cos \omega_r t \text{ and } a_r \cos (\omega_r t + \phi_r)$$

The correlation coefficient of these two is

$$\psi = 2 \; . \; \overline{\cos \omega_r t \; . \; \cos (\omega_r t + \phi_r)}$$
$$= \cos \phi_r \qquad (2.16)$$

so that if the phase difference is $\pi/2$ rad there is actually no correlation. But there may be values of ϕ_r at which the correlation is again unity. If the bandwidth of the noise (as received and restricted by the transducer) is very narrow, the correlation will change more or less as for the single frequency just considered; but if it is rather wider, the correlation will never reach unity at any direction other than the normal. It can be shown[13] that if

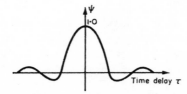

FIG. 2.7. Form of correlation function.

correlation coefficient is plotted against time difference, a curve as shown in Fig. 2.7 is obtained. This curve is called the correlation function. For an infinite number of noise components ($n = \infty$ in eqn. (2.7)), the curve has the form $[(\sin x)/x] \cos \omega_0 \tau$, where $x = b\tau$, $\omega_0 = $ midband frequency, $2b = $ bandwidth in rad/sec, and τ is the time difference, or delay. The function is usually written $\psi(\tau)$. Its maximum value is unity only when the two waveforms are identical; in general, the peak value lies between 0 and 1.

The correlation coefficients and functions which have been discussed above are between two different noise waveforms or two different channels. They are therefore called "cross-correlation" factors and functions. Another kind of correlation function is often important, the "autocorrelation function". This is the $\psi(\tau)$ obtained when the correlation is taken between the waveform and itself delayed by τ. Knowledge of this function enables the frequency spectrum of the noise to be determined and is useful in calculations rather more advanced than those involved in this book.

5. Signal Processing and Display

In any underwater acoustic system (as in many other kinds of system too) there is some particular form of information which is wanted. The signals corresponding to this have to be identified and displayed or presented to the operator(s). In a sonar system, for example, it is the echo-signal from the target which has to be identified in order that information regarding range, bearing, size, Doppler shift, etc. may be obtained. In a communication system it is the modulated speech signals which have to be identified. Noise and other irrelevant signals have to be rejected as far as possible. It is usual to regard the part of the system which is concerned with identifying and separating the wanted signals and rejecting the noise as having the function of "signal processing".

As a simple illustration, suppose the wanted signal is just a sinewave and the information is indicated by its amplitude (which is constant over a long period of time). The noise accompanying this signal in the medium may, in general, be so great as to prevent an accurate measurement of the signal being made—or even so great as to prevent its presence being noticed. But if the receiver includes a bandpass filter of very narrow bandwidth, most of the noise can be rejected (since it covers a wide spectrum) although the signal, which being a constant tone, needs strictly only an infinitesimal bandwidth, will be passed. The signal can then be accurately determined.

In this simple example the signal required only an infinitesimal bandwidth. But as we saw in Section 2, this restriction of band means that the time required for any change in amplitude is infinite, and thus no further information can be transmitted by the signal. The information capacity of the system is infinitesimal. If information is to be conveyed by variations in the amplitude or the frequency of the signal (as in a communication system) or by the time of occurrence of echo-pulses (as in a sonar system), then clearly a finite bandwidth is necessary. But to observe this information with greatest accuracy it is necessary to exclude as much noise as possible and keep the bandwidth as small as possible. There is, obviously, for any particular specified signal, an optimum receiver response. This leads quite naturally to the idea of a filter being "matched" to the signal. The general mathematical development of this idea can be found in books on information theory. We shall discuss it here only in a qualitative way and in respect to a simple sonar system.

Suppose in the sonar system the duration of the transmitted pulse is T_p sec. This can be chosen to give a specified resolution in range as discussed in Section 2. Then the transducers and the receiving amplifier, etc., must have a bandwidth of about $1/T_p$ c/s to permit a proper response to the echo pulses and yet accept as little noise as possible. If moving targets are expected, a somewhat larger bandwidth is necessary, to allow for the Doppler shift of frequency (which is presumably not known in advance). After amplification the signal has to be rectified so that the envelope of the pulses is obtained. This rectification process, being nonlinear, introduces complications since the signal and noise interact. If the signal-to-noise ratio is marginal or poor (say, around or below unity), the actual law of the rectifier has little effect on the signal-to-noise performance,[14] which is then that corresponding to a square-law rectifier, and gives a signal-to-noise ratio† (R_0) at the output (usually called the "video" part of the system by analogy with television) which is equal to the square of the input ratio R_1, i.e.

$$R_0 = R_1^2 \qquad (2.17)$$

If single echo-pulses are to be examined or displayed, then this is their signal-to-noise ratio, and detection and identification may be very difficult.

However, the transmitter emits pulses regularly every T_r sec.

† Defined as the ratio of the change in output d.c. level when the signal pulse is applied to the noise level when the signal is absent.

It is known therefore that the echo-pulses may be received repetitively at this interval provided there is no relative motion of the target along the sound beam. If by some means the signals returned from n successive emissions can be stored, delayed by multiples of T_r sec, and added together, then the echo amplitude will be increased by a factor of n. Now the noise accompanying the echo-pulses will be uncorrelated between one emission period and another (since it is random in time anyway), and this will be true (very largely) for most other kinds of background, such as reverberation. Thus the n noise voltages added together will give a resultant noise voltage increased by a factor of only $\sqrt{n}$. The signal-to-noise ratio after this process of signal integration (or correlation) is therefore

$$R_{0n} = \sqrt{(n)} \, R_0 = \sqrt{(n)} \, R_1^2 \qquad (2.18)$$

Another way of looking at this result is that if a particular value of output signal-to-noise ratio gives a just-detectable signal (e.g. it may give a 50% probability of the echo being correctly identified), then the value of R_1 corresponding to this output is called the "detection threshold". Then clearly, if n repetitions of the signal are integrated, the threshold is improved (i.e. lowered) by a factor of $n^{1/4}$. This is usually expressed as a threshold improvement of "$1 \cdot 5$ dB per doubling" of the number of repetitions.

Returning now to the idea of the matched filter, we see that the signal envelope waveform consists of a pulse pattern repeating every T_r sec. Its frequency spectrum, therefore, consists of a series of lines spaced $1/T_r$ c/s apart, as shown at (a) in Fig. 2.8. If a filter is used after the rectifier which has a series of very narrow pass bands corresponding exactly to these spectral lines, as shown at (b) in Fig. 2.8, then all the signal will be accepted, but the noise will be rejected. This is clearly a filter matched to the signal in a very close way. It is usually referred to as a "comb" filter. Since, in general, the echo pattern is not held constant indefinitely, and we may say that only n repetitions of an identical pattern occur, there will be a small spectrum spread around each of the lines mentioned above, the width of this spread being inversely proportional to n. Thus the smaller is n, the wider must be the "teeth" of the comb, and the smaller, therefore, the improvement in signal-to-noise ratio. It can be shown that this effect is identical with the integration of n repetitions by storage as discussed above, and the latter is, of course, the more practicable method of realization.

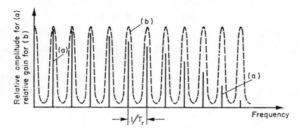

FIG. 2.8. Showing spectrum of output signal (a) in a pulsed sonar system, and the response of filter circuit (b) for improving signal-to-noise ratio. (N.B. For clarity a much smaller value of ratio (repetition period–pulse duration) has been assumed than would occur in a practical system.)

The interesting thing about the chemical recorder type of display, as used in a sonar system, is that it performs a rather similar function to the integration method discussed above. The signals received from successive emissions are stored as the visible marks on each trace, and successive traces are drawn side by side. The human eye (and brain) can then observe the regular repetitions of genuine echo-signals among the non-repetitive noise, and detection and identification is improved as the number of traces increases. Actually, the detection threshold improves, as the number of traces increases, rather more rapidly than the "1·5 dB per doubling" law quoted above; the figure is nearer 2·4 dB. Full investigations of this effect have been reported in the literature.[15]

If we turn to communication and telemetry systems, we find the problems of signal processing are fully covered in numerous textbooks on communications engineering,[9,10] and as the problems are so little related to the specifically underwater acoustic aspects of the systems, it does not seem desirable to pursue them further in this book.

References

1. WOODWARD, P. M., *Probability and Information Theory, with Applications to Radar*, Pergamon (1953).
2. BELL, D. A., *Information Theory and its Engineering Applications*, Pitman (1962).
3. GOLDMAN, S., *Frequency Analysis, Modulation and Noise*, McGraw-Hill, New York (1948).
4. GOLDMAN, S., *Information Theory*, Prentice Hall, London (1953).
5. GUILLEMIN, E., *Communication Networks*, Vol. II, Wiley, New York (first published 1931).
6. TUCKER, D. G., *Elementary Electrical Network Theory*, Pergamon, Oxford (1964). (See p. 155.)
7. TUCKER, D. G., Bandwidth and Speed of Build-up as Performance Criteria in Pulse and Television Amplifiers, *J. Inst. Elect. Engrs.* **94**, 218 (1947).
8. TUCKER, D. G., WELSBY, V. G., and KENDELL, R., Electronic Sector Scanning, *J. Brit. Inst. Radio Engrs.* **18**, 465 (1958).
9. BLACK, H. S., *Modulation Theory*, Van Nostrand, New York (1953).
10. BROWN, J., and GLAZIER, E. V. D., *Principles of Telecommunication*, Chapman and Hall, London (1964).
11. See, e.g., VAN DER ZIEL, A., *Noise*, Chapman and Hall, London (1955).
12. SLACK, M., The Probability Distributions of Sinusoidal Oscillations Combined in Random Phase, *J. Inst. Elect. Engrs.* **93**, Part III, 76 (1946).
13. NODTVEDT, H., The Correlation Function in the Analysis of Directive Wave Propagation, *Phil. Mag.* **41**, 1022 (1951).
14. BENNETT, W. R., Response of a Linear Rectifier to Signal and Noise, *J. Acoust. Soc. Amer.* **15**, 164 (1944); also numerous later papers, e.g. TUCKER, D. G., and GRIFFITHS, J. W. R., Detection of Pulse Signals in Noise, *Wireless Engineer* **30**, 264 (1953).
15. TUCKER, D. G., Detection of Pulse Signals in Noise: Trace-to-Trace Correlation in Visual Displays, *J. Brit. Inst. Radio Engrs.* **17**, 319 (1957).

CHAPTER 3

The Propagation of Sound

1. Introduction

Acoustic radiation appears to an observer as a time-varying change of pressure which travels progressively through an acoustic medium. This pressure change, which is equal to the difference between the ambient and instantaneous values, is known as the *excess* or *acoustic pressure*.

As it propagates, the acoustic pressure produces compressions and rarefactions in successive elements of the medium along its path and particles within these elements are displaced; because sound is propagated in this manner, it follows that all acoustic media must be *compressible*. A local disturbance cannot therefore be transmitted instantaneously throughout the medium and sound must travel with a finite velocity depending upon the *compressibility* and *density* of the medium.

The time waveform of the acoustic pressure depends largely upon its method of generation. When generated by an impulsive source, such as an explosion, it often takes the form of a unidirectional pressure pulse. The more usual time waveform, and the one with which we shall mainly be concerned, however, consists of a series of alternate compressions and rarefactions forming a *simple harmonic* or sinusoidal time function.

The corresponding particle displacement within each element of the medium is simple harmonic also and the acoustic pressure is propagated as a sinusoidal *wave*.

As this wave propagates successive elements of the medium are compressed and then released. The medium, being compressible, exerts a restoring force which accelerates each particle in each element and restores them to their undisturbed positions. By this time, the restoring force has fallen to zero, but owing to the *inertia* of the medium, the particles are carried past their undisturbed positions and in a loss-less medium make excursions of equal amplitude on the other side of their datum. This process is repeated continuously as the wave is propagated. If the medium absorbs some of the sound energy, successive excursions will be of decreasing amplitude and the wave is said to be *attenuated*. The various causes of absorption arising in fluid media will be discussed later in this chapter.

2. Types of Acoustic Wave

A number of different types of wave motion can be used for the transmission of acoustic energy. These fall into two categories, those which propagate with equal facility in all directions, *homogeneous waves;* and *inhomogeneous waves*, which propagate readily in one plane but are severely attenuated in any direction normal to this plane. Inhomogeneous waves, being confined to a single plane, are usually only found on the surface layer of an acoustic medium and for this reason are often referred to as *surface waves*.

Rayleigh waves and Love waves are both examples of inhomogeneous waves. In the former, which is similar to the wave motion at the surface of the sea, particles of the medium describe elliptical orbits, as shown in Fig. 3.1(a). Because of the compressibility of the medium successive particles along the direction of propagation reach the same point on their orbits at different times and a wave is apparently propagated along the surface. Love waves, as shown in Fig. 3.1(b), are found only in thin layered media bounded by a fluid layer on one side and a solid layer on the other, both of which are sufficiently extensive to allow reflections from their other boundary to be neglected. Within this thin layer the particle motion is entirely horizontal and normal to the direction of propagation.

Rayleigh and Love waves, with frequencies in the range 0·1–

10 c/s are observed in the uppermost layers of the sea-floor. Owing to their low frequencies they suffer very little absorption and can propagate over considerable distances; they are, in consequence of great importance in seismological studies.

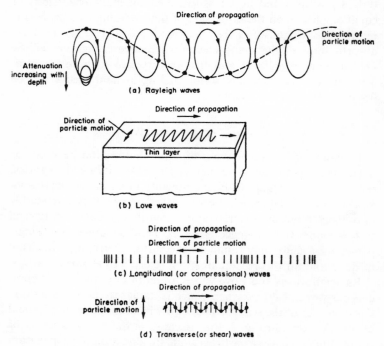

FIG. 3.1. Types of acoustic wave motion.

There are two types of homogeneous wave, *longitudinal or compressional* waves and *transverse or shear waves*. With these, the particle motion is along and perpendicular to the direction of propagation respectively, as shown in Fig. 3.1(c) and (d).

Transverse waves set up large shear stresses within the medium; consequently, only solids and certain liquids possessing extremely high viscosities can sustain these stresses without shearing. Shear waves cannot therefore propagate through gaseous or the majority of fluid media including water.

Longitudinal waves are by far the most important type of wave in acoustics, in general, and this is particularly so underwater; consequently, all the analysis in this and succeeding chapters will be developed specifically for this type of wave. We shall also confine our attention to simple harmonic waves.

As a simple harmonic wave propagates, the magnitude of the acoustic disturbance, however measured, varies sinusoidally in time at every part in the medium. The spatial distribution of the disturbance, measured at any fixed time is sinusoidal also. The surface joining regions within the medium undergoing the same amount of perturbation during the same compression or rarefaction cycle is known as a *wavefront* and its shape enables the classification of acoustic waves to be subdivided further.

Waves generated by a point source (or one whose dimensions are small compared with their wavelength) in an homogeneous medium propagate with spherical symmetry, their wavefronts are spherical and they are termed *spherical waves*. If the medium is bounded by two plane parallel boundaries (as in the case of the sea) waves generated by a point source spread with circular symmetry only in the horizontal plane. Their wavefronts are cylindrical and they are known as *cylindrical waves*. If the source is an infinite plane surface, the resulting wavefronts are also plane, no spreading occurs and the waves are known as *plane waves*.

Although true plane waves cannot be generated in practice, both spherical and cylindrical waves approximate to plane waves when they are sufficiently removed from their sources.

3. Elastic Properties of Fluid Media

An acoustic pressure wave applies a stress to successive elements of the medium through which it propagates. The resulting particle motion in each element is determined by the mechanical properties of the medium, in particular, its density ρ and the elastic modulus describing the difficulty with which it is compressed. In a solid, this elastic modulus is dependent upon the orientation of the medium relative to the acoustic wave, and in a completely anisotropic solid, thirty-six different elastic constants, of which twenty-one are independent, are required to specify completely the stress–strain relationships within it.

Fluids and gases, being isotropic, require only one such elastic constant, the *compressibility s* measured in m^2/newton which is defined as the volumetric strain produced per unit applied stress:

$$s = \frac{\Delta v}{v_0}\bigg/ p$$

where Δv is the change in the original volume v_0 on application of an excess hydrostatic pressure p measured in newtons/m^2.

The reciprocal of the compressibility is known as the *volume elasticity* or *bulk modulus* κ measured in newtons/m^2

$$\kappa = \frac{1}{s} = p\bigg/ \frac{\Delta v}{v_0} \qquad (3.1)$$

and is often used in preference to the compressibility in expressions relating acoustic quantities.

For the purposes of an elementary treatment of acoustic theory, the bulk modulus is assumed to be constant. In practice, however, as the excess pressure amplitude is increased it is observed that the volumetric strain fails to increase in proportion and the bulk modulus increases considerably above its low-amplitude value. The elementary theory, therefore, is only applicable to low-amplitude acoustic waves.

The passage of an acoustic wave through the medium can be characterized by certain parameters which vary periodically with both space and time. It is useful to consider a few of these in detail and derive some basic relationships between them.

Particle Displacement, ξ, is the amount of displacement of a particle from its mean position within the medium, measured in metres, under the action of the acoustic pressure.

Particle Velocity, *u*, is the velocity of a particle in the medium, given by the time derivative of the particle displacement and measured in m/sec; $u = \mathrm{d}\xi/\mathrm{d}t$.

Particle Acceleration, *a*, is the time derivative of particle velocity, measured in m/sec²; $a = \mathrm{d}u/\mathrm{d}t = \mathrm{d}^2\xi/\mathrm{d}t^2$.

Acoustic or Excess Pressure, *p*, is the change of pressure from the mean value. It is equal to the difference between the instantaneous pressure *P* and the hydrostatic pressure P_0; $p = P - P_0$ measured in newtons/m².

Condensation, *S*, is the fractional change in density resulting from the action of the acoustic pressure, defined as

$$S = (\rho - \rho_0)/\rho_0 \tag{3.2}$$

where ρ and ρ_0 are the instantaneous and mean densities respectively. The mass of the medium must remain unaltered during the passage of the acoustic wave, and consequently eqn. (3.2) can be written as

$$S \simeq -\Delta v/v_0 \tag{3.3}$$

Combined with eqn. (3.1) this gives an equation relating bulk modulus and excess pressure:

$$p = \kappa S \tag{3.4}$$

Equation (3.2) can be written conveniently in the form

$$\rho = \rho_0(1 + S)$$

whence

$$\frac{\mathrm{d}\rho}{\mathrm{d}t} = \rho_0 \frac{\mathrm{d}S}{\mathrm{d}t} \tag{3.5}$$

Velocity Potential, ϕ, is related to the three components of particle velocity *u*, *v* and *w*, along the *x*-, *y*- and *z*-directions respectively, by the expressions

$$u = \frac{\mathrm{d}\phi}{\mathrm{d}x}, \quad v = \frac{\mathrm{d}\phi}{\mathrm{d}y}, \quad w = \frac{\mathrm{d}\phi}{\mathrm{d}z}$$

Although an entirely theoretical concept, the fact that it is a scalar quantity makes it useful; several works on acoustics make repeated use of it, but because it has no physical significance it will not be used in this book.

Propagation Velocity, c, is the velocity with which the acoustic wave passes through the fluid medium. It is related to the mechanical constants of the medium, as will be seen in Section 4, by

$$c = \sqrt{(\kappa/\rho_0)} \qquad (3.6)$$

In the derivation of this equation no account is taken of the thermodynamical processes involved as the acoustic wave is propagated, so the value of κ in eqn. (3.6) is the value measured under isothermal conditions which presuppose that the compressions and rarefactions associated with the acoustic wave take place at constant temperature. In reality, these isothermal conditions do not apply. During the compression cycle each elemental volume in turn is raised in temperature; the actual temperature gradient formed, however, is small owing to the low excess pressures involved. The duration of the compression is usually insufficient for any significant heat flow to take place before the compression is relaxed, consequently, the thermodynamical process is far from isothermal and is said to be *adiabatic*; eqn. (3.6) must be modified accordingly.

As the first step in this modification, eqn. (3.6) can be written in a more general form which is often more convenient by using eqn. (3.1) thus:

$$c = \sqrt{(\mathrm{d}p/\mathrm{d}\rho)} \qquad (3.7)$$

Under isothermal and adiabatic conditions the excess pressure and density are related by

$$p = A(\rho - \rho_0) \quad \text{and} \quad p = A'(\rho - \rho_0)^\gamma \qquad (3.8)$$

respectively, where γ is the ratio of the specific heats, C_p and C_v of the medium, measured under conditions of constant pressure and constant volume respectively.

The isothermal and adiabatic propagation velocities as given by eqn. (3.7) are

$$c_{IS}^2 = \left(\frac{dp}{d\rho}\right)_{IS} = A = p/(\rho - \rho_0)$$

and

$$c_{AD}^2 = \left(\frac{dp}{d\rho}\right)_{AD} = A'\gamma(\rho - \rho_0)^{\gamma-1} = \gamma p/(\rho - \rho_0)$$

from which it can be seen that

$$c_{AD}^2 = \gamma c_{IS}^2$$

Combining this expression with eqn. (3.6) gives the velocity of propagation under real (adiabatic) conditions as

$$c = \sqrt{(\gamma\kappa/\rho)} \tag{3.9}$$

For sea-water at 13°C, for which $\gamma = 1\cdot01$, $\kappa = 2\cdot28 \times 10^9$ newtons/m² and $\rho = 1026$ kg/m³, this expression gives the propagation velocity as 1500 m/sec.

This value is only correct at the surface of the sea and where the salinity is thirty-five parts per thousand. At any other depth, temperature or salinity the propagation is given by the empirical equation given in Chapter 1.

If the sea is moving with a component of velocity along the direction of propagation the effective velocity of propagation is increased or decreased accordingly. Under these conditions the frequency of the acoustic signal remains constant, however, so the change in propagation velocity results in a change in wavelength.

If the sound source is moving relative to a fixed receiver with a component of velocity along the line joining the two of v_s (here a positive value of v_s denotes movement in the direction of propagation) the wavelength must alter. The observed frequency at the receiver f_r is then related to the source frequency f_s by

$$f_r = f_s \left[\frac{c}{c - v_s}\right]$$

i.e. it increases if the source moves towards the receiver and decreases if it moves away.

When the receiver alone is moving the wavelength remains unaltered but the apparent velocity of propagation is changed. If the velocity of the receiver relative to the source is v_r (the sign convention adopted above is again used) the apparent velocity of propagation is $c - v_r$ hence the observed frequency at the receiver

$$f_r = f_s \left[\frac{c - v_r}{c} \right]$$

The general case, when both source and receiver are moving, is covered by the expression

$$f_r = f_s \left[\frac{c - v_r}{c - v_s} \right]$$

This frequency change resulting from relative motion of source and receiver is known as the *Doppler effect* and the actual change of frequency

$$f = f_r - f_s = f_s \left[\frac{c - v_r}{c - v_s} - 1 \right]$$

is termed the *Doppler shift*.

4. The General Acoustic Wave Equation for a Wave Propagating in an Infinite Loss-less Homogeneous Medium

A wave equation relates the space and time derivatives of the various acoustic parameters (acoustic pressure, etc.) describing the wave motion as it propagates. Its solution, using specific boundary conditions, provides exact equations for the propagating wave and is therefore of considerable importance. It is derived by applying the principle of the conservation of mass and Newton's second law (force = mass × acceleration) to an elemental volume of the medium. For this derivation, it is proposed to adhere to the assumption regarding small amplitude vibrations mentioned previously.

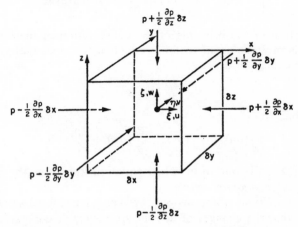

FIG. 3.2. Pressures acting on an elemental fluid volume.

Consider the element of the medium $\delta x \delta y \delta z$, shown in Fig. 3.2 which is subject to an acoustic wave propagating in a general direction. This wave will cause particle motion at the centre of the element with amplitudes ξ, η and ζ and velocity components u, v and w along the x-, y- and z-axes respectively.

The mass of material entering one yz face of this element in time δt is

$$\rho(u\delta t)\,\delta y \delta z$$

and the mass leaving the opposite face is

$$\left[\rho u + \frac{\partial}{\partial x}(\rho u)\,\delta x \right] \delta t \delta y \delta z$$

The excess mass of material leaving the element is therefore

$$\frac{\partial}{\partial x}(\rho u)\,\delta x \,.\, \delta t \delta y \delta z$$

The same argument can be applied to the other faces also to give
the total mass leaving the element as

$$\left[\frac{\partial}{\partial x}(\rho u) + \frac{\partial}{\partial y}(\rho v) + \frac{\partial}{\partial z}(\rho w)\right]\delta t\delta x\delta y\delta z$$

Application of the principle of the conservation of mass shows
that this must be equal to the product of the elemental volume
and the change of density

$$-\frac{\partial \rho}{\partial t}\,\delta t\delta x\delta y\delta z.$$

Therefore

$$\frac{\partial}{\partial x}(\rho u) + \frac{\partial}{\partial y}(\rho v) + \frac{\partial}{\partial z}(\rho w) + \frac{\partial \rho}{\partial t} = 0$$

This equation is one form of the equation known as the *hydro-
dynamical equation of continuity*.

Low-amplitude vibrations only are being considered so we can
assume that the changes of density are negligible compared with
the changes in particle velocity. Thus

$$\frac{\partial}{\partial x}(\rho u) = \rho_0\frac{\partial u}{\partial x}, \; \frac{\partial}{\partial y}(\rho v) = \rho_0\frac{\partial v}{\partial y} \text{ and } \frac{\partial}{\partial z}(\rho w) = \rho_0\frac{\partial w}{\partial z}$$

where ρ_0 is the density of the undisturbed element.

Using these relationships and eqn. (3.5) the equation of continuity can be expressed in its more usual form

$$\frac{\partial u}{\partial x} + \frac{\partial v}{\partial y} + \frac{\partial w}{\partial z} + \frac{\partial S}{\partial t} = 0 \tag{3.11}$$

Newton's second law is applied by equating the differences in
external pressure acting upon each pair of opposite faces of the
elemental volume to the product of its mass and the acceleration
along the x-, y- and z-axes. If the pressure at the centre of the
element is p, the pressures on the opposite yz-faces are

$$p - \tfrac{1}{2}\frac{\partial p}{\partial x}\delta x \text{ and } p + \tfrac{1}{2}\frac{\partial p}{\partial x}\delta x$$

therefore

$$-\frac{\partial p}{\partial x}\,\delta x \delta y \delta z = \rho_0 \delta x \delta y \delta z\,\frac{\partial u}{\partial t}$$

i.e.

$$-\frac{\partial p}{\partial x} = \rho_0\,\frac{\partial u}{\partial t} \tag{3.12}$$

Similarly, by considering the forces on the other faces

$$-\frac{\partial p}{\partial y} = \rho_0\,\frac{\partial v}{\partial t};\ -\frac{\partial p}{\partial z} = \rho_0\,\frac{\partial w}{\partial t} \tag{3.12a}$$

If these equations are partially differentiated with respect to x, y and z respectively and added

$$-\left(\frac{\partial^2 p}{\partial x^2} + \frac{\partial^2 p}{\partial y^2} + \frac{\partial^2 p}{\partial z^2}\right) = \rho_0\,\frac{\partial}{\partial t}\left(\frac{\partial u}{\partial x} + \frac{\partial v}{\partial y} + \frac{\partial w}{\partial z}\right)$$

This equation can be simplified using eqn. (3.11) to

$$\frac{\partial^2 p}{\partial x^2} + \frac{\partial^2 p}{\partial y^2} + \frac{\partial^2 p}{\partial z^2} = \rho_0\,\frac{\partial^2 S}{\partial t^2} \tag{3.13}$$

and expressed in a more acceptable form by using the second derivative of eqn. (3.4) with respect to t, to give

$$\frac{\kappa}{\rho_0}\left(\frac{\partial^2 p}{\partial x^2} + \frac{\partial^2 p}{\partial y^2} + \frac{\partial^2 p}{\partial z^2}\right) = \frac{\partial^2 p}{\partial t^2}$$

It is usually written in terms of the Laplacian Operator

$$\nabla^2 = \frac{\partial^2}{\partial x^2} + \frac{\partial^2}{\partial y^2} + \frac{\partial^2}{\partial z^2}\ \text{and thus becomes}$$

$$\frac{\partial^2 p}{\partial t^2} = \frac{\kappa}{\rho_0}\,\nabla^2 p \tag{3.14}$$

It is clear that this wave equation relates the variation of pressure along various directions in space, as seen at a particular instant of time, to the variation with time which occurs at any particular point in space. It is entirely reasonable that the parameter on which is based the relationship between the spatial and temporal variations should be the propagation velocity of the wave, i.e. $c^2 = \kappa/\rho_0$, from equation (3.6).

The wave equation can also be expressed in terms of the other acoustic parameters, particle velocity, acceleration, etc., but it applies only to waves propagating in an infinite homogeneous medium in which no absorption occurs.

Equation (3.14) is expressed in terms of Cartesian coordinates but it can be used more conveniently to describe acoustic radiation from point or spherical sources, under conditions of spherical or cylindrical spreading, simply by expressing the Laplacian operator in terms of the appropriate coordinate system, spherical or cylindrical as shown in Section 6.

5. Plane Waves in an Infinite Homogeneous Medium

When the acoustic pressure is independent of both y- and z-coordinates, i.e. it acts only in the x-direction, eqn. (3.14) reduces to a simple one-dimensional expression describing the propagation of plane waves.

$$\frac{\partial^2 p}{\partial t^2} = c^2 \frac{\partial^2 p}{\partial x^2} \qquad (3.14a)$$

This equation has a general solution of the form:

$$p = F_1(ct - x) + F_2(ct + x)$$

which represents two plane acoustic disturbances travelling in the positive and negative x-directions respectively each with a velocity c. When propagation of acoustic waves through an infinite medium is under consideration, one of these solutions may be ignored, as in eqn. (3.21), since it represents a reflection of the other from a plane boundary placed normal to the direction of propagation.

Only the simple harmonic form of solution will be treated here and this can be expressed in one of two ways, either in complex form:

$$p = A \exp [\text{j}(\omega t - kx)] + B \exp [\text{j}(\omega t + kx)] \qquad (3.15)$$

which is a true solution but does not truly represent the real physical wave; or in real terms which correspond to the real parts of the complex expression:

$$p = A \cos (\omega t - kx) + B \cos (\omega t + kx) \qquad (3.16)$$

In both of these equations, A and B represent the pressure amplitudes of the two plane waves of angular frequency ω travelling in opposite directions and k, defined by

$$k = \omega/c = 2\pi/\lambda \qquad (3.17)$$

is a wavelength constant referred to as the *acoustic wave number*.

The complex form of solution is often more convenient to use in mathematical manipulations, and is, in fact, more frequently encountered in the literature than the cosine form; for this reason we shall use it here. But there are dangers in its use, since the physical wave is represented truly only by the real part of the complex function; so long as the system is linear in its propagation–amplitude response, this does not matter; but once non-linear response is encountered, the complex form must be abandoned.

For the present only waves travelling in the positive x-direction will be considered.

Plane waves are independent of the y- and z-coordinates, and consequently the three components of eqn. (3.12) reduce to a single equation

$$-\frac{\partial p}{\partial x} = \rho_0 \frac{\partial u}{\partial t} \qquad (3.18)$$

from which the particle acceleration is

$$a = -\frac{1}{\rho_0} \frac{\partial p}{\partial x} \qquad (3.19)$$

Integration of this equation with respect to time gives the instantaneous particle velocity

$$u = -\frac{1}{\rho_0} \int \frac{\partial p}{\partial x} \, dt \tag{3.20}$$

and enables the coefficient A in the equation for acoustic pressure

$$p = A \exp [j (\omega t - kx)] \tag{3.21}$$

to be determined. Differentiation of eqn. (3.21) with respect to x and integration with respect to time gives

$$\int \frac{\partial p}{\partial x} \, dt = -\frac{k}{\omega} A \exp [j (\omega t - kx)] \tag{3.22}$$

Using eqns. (3.20) and (3.17) the instantaneous particle velocity is therefore:

$$u = \frac{A}{\rho_0 c} \exp [j (\omega t - kx)]$$

$$= \hat{u} \exp [j (\omega t - kx)] \tag{3.23}$$

where $\hat{u} = A/\rho_0 c$ is the peak particle velocity; therefore the coefficient in eqn. (3.21) is given by

$$A = \rho_0 c \hat{u}$$

and the expression for acoustic pressure becomes

$$p = \rho_0 c \hat{u} \exp [j (\omega t - kx)] \tag{3.24}$$

which has a peak value $\hat{p} = \rho_0 c \hat{u}$.

The particle displacement in a plane wave can be obtained by further integration of eqn. (3.20) thus

$$\xi = -\frac{1}{\rho_0} \int \int \frac{\partial p}{\partial x} \, dt dt'$$

which on applying eqn. (3.22) can be put in the form:

$$\xi = \frac{1}{\omega^2 \rho_0} \cdot \frac{\partial p}{\partial x}$$

The general equation of continuity may be expressed in yet

another form if eqn. (3.11) is integrated with respect to time:

$$\frac{\partial \xi}{\partial x} + \frac{\partial \eta}{\partial y} + \frac{\partial \zeta}{\partial z} = -S$$

This expression also gives a general equation for the condensation at any point in the medium. The condensation due to a plane wave can be obtained from this expression simply by ignoring the partial derivatives with respect to y and z, thus:

$$\frac{\partial \xi}{\partial x} = -S \qquad (3.25)$$

Finally, a useful expression relating acoustic pressure and condensation may be obtained by combining eqns. (3.4) and (3.6), thus:

$$p = \rho_0 c^2 S \qquad (3.26)$$

5.1. *Energy Density and Acoustic Intensity of a Plane Wave*

The energy of a plane wave comprises two components: (a) the *kinetic energy* associated with particle motion, and (b) the *potential energy* stored during compression and rarefaction of successive fluid elements. There is an interchange of energy between these two forms, the maximum of one being coincident with the minimum of the other.

The kinetic energy stored in a fluid element of volume v whose length is chosen so that the particle velocity u can be considered uniform throughout the element is

$$\text{K.E.} = \tfrac{1}{2} \rho_0 v_0 u^2 \qquad (3.27)$$

The potential energy is associated with the change in elemental volume $\mathrm{d}v$ under the action of the acoustic pressure p, an increase in pressure resulting in a decrease in volume so giving rise to a negative sign in the potential energy expression

$$\text{P.E.} = -\int p \, \mathrm{d}v \qquad (3.28)$$

The volume of the fluid element is related to the external acoustic pressure by eqns. (3.3) and (3.26) thus

$$v = v_0 \left(1 - p/\rho_0 c^2\right)$$

where v_0 is the volume in the absence of acoustic pressure.
Thus

$$\mathrm{d}v = -\frac{v_0 \, \mathrm{d}p}{\rho_0 c^2}$$

Equation (3.28) can be rewritten therefore as

$$\mathrm{P.E.} = \int p \cdot \frac{v_0 \mathrm{d}p}{\rho_0 c^2}$$

$$= \frac{1}{2} \frac{p^2}{\rho_0 c^2} v_0$$

The total instantaneous energy of the acoustic waves in the fluid element is equal to the sum of the kinetic and potential energies

$$E = \mathrm{K.E.} + \mathrm{P.E.} = \tfrac{1}{2} \rho_0 \left(u^2 + \frac{p^2}{\rho_0^2 c^2}\right) v_0$$

from which may be obtained an expression for the *instantaneous energy density* measured in joules/m³:

$$\mathscr{E} = E/v_0 = \tfrac{1}{2} \rho_0 \left(u^2 + \frac{p^2}{\rho_0^2 c^2}\right) \tag{3.29}$$

The pressure and velocity of a plane wave are, however, related by the expression

$$p = \rho_0 c u$$

therefore eqn. (3.29) can be simplified to

$$\mathscr{E} = \rho_0 u^2 \tag{3.30}$$

The particle velocity, u, in this expression is a function of time, however, and the time average of eqn. (3.30) must be sought. This time average is given by

$$\bar{\mathscr{E}} = \frac{1}{T} \int\limits_0^T \mathscr{E} \, \mathrm{d}t$$

as $T \to \infty$ in general; but for a simple harmonic wave T may be $2\pi/\omega$. Using the real part only of eqn. (3.23) for this purpose, $u = \hat{u} \cos (\omega t - kx)$, the mean energy density $\bar{\mathscr{E}}$ is therefore given by

$$\bar{\mathscr{E}} = \frac{\rho_0 \hat{u}^2}{T} \int\limits_0^T \cos^2 (\omega t - kx) \, dt$$

$$= \frac{\rho_0 u^2}{2T} \int\limits_0^T [1 + \cos 2 (\omega t - kx)] \, dt$$

$$= \tfrac{1}{2} \rho_0 \hat{u}^2 \tag{3.31}$$

The mean value is therefore half the maximum value.

The *acoustic intensity*, I, of an acoustic wave is defined as the mean rate of flow of energy through a unit area normal to the direction of propagation. If the propagation velocity is c, the mean energy contained in a column of unit area and length $c\,dt$ is

$$E = \bar{\mathscr{E}} \, c\,dt$$

from which can be obtained the rate of energy flow or acoustic intensity, measured in W/m²

$$I = \frac{dE}{dt} = \bar{\mathscr{E}} c = \tfrac{1}{2} \rho_0 c \hat{u}^2$$

This relationship can be expressed in alternative forms, by using eqn. (3.24), as

$$I = \tfrac{1}{2} \rho_0 c \hat{u}^2 = \tfrac{1}{2} \hat{p} \hat{u} = \tfrac{1}{2} \frac{\hat{p}^2}{\rho_0 c} \tag{3.32}$$

At this point it is convenient to draw certain analogies between acoustical and electrical quantities. Acoustic intensity measured on a unit area basis and electrical power are equivalent; therefore a comparison may be made between the relationships of eqn. (3.32) and the simple a.c. electrical relationships between voltage, current and impedance. This shows that particle velocity and pressure are analogous to current and voltage respectively. The quantity $\rho_0 c$ is a measure of the acoustical behaviour of the medium and is analogous to electrical impedance; it is therefore referred to as the *characteristic acoustic impedance*.

Electrical quantities are often measured as root-mean-square (r.m.s.) values, where the r.m.s. value is equal to the peak value divided by $\sqrt{2}$. In the same way it is often convenient to measure acoustic pressure and particle velocity as r.m.s. values and the expressions of eqn. (3.32) become

$$I = \rho_0 c\, u_{rms}^2 = p_{rms} u_{rms} = \frac{p_{rms}^2}{\rho_0 c} \qquad (3.33)$$

where $u_{rms} = \hat{u}/\sqrt{2}$ and $p_{rms} = \hat{p}/\sqrt{2}$

5.2. *Specific Acoustic Impedance*

This is the actual impedance presented by the medium to an acoustic source or receiver immersed within it. We shall see later that not only is it dependent upon the physical properties of the medium but it is a function of the particular type of wave propagated also.

It is defined as the ratio of acoustic pressure to the associated particle velocity, thus

$$Z = p/u \qquad (3.34)$$

and in general p and u will not be in the same phase, so that they (along with Z) have to be considered as *phasors*.† Actually, u is also a vector in the true spatial sense, but it is its phasor properties that we are concerned with here. Thus

$$Z = R + \mathrm{j}X$$

where R is termed the *specific acoustic resistance* and X, the *specific acoustic reactance*.

For plane progressive waves the phase of p and u is identical;

† For a discussion on phasors, which are indicated by bold type, see any elementary book on electrical network theory (e.g. Tucker[1]). They are treated like vectors.

thus

$$Z = \frac{p}{u} = \frac{\hat{p} \exp(-jkx)}{\hat{u} \exp(-jkx)} = \frac{\hat{p}}{\hat{u}} = \frac{\rho_0 c u}{u}$$

i.e. it is an entirely real quantity:

$$Z = \rho_0 c \tag{3.35}$$

The product $\rho_0 c$ is encountered in many expressions relating acoustic quantities and has far more significance then either ρ_0 or c taken individually. For this reason it is given a separate identity termed the characteristic acoustic impedance (although it is strictly a resistance) measured in units of kg/m² sec which are sometimes, but not often, referred to as *rayls* in honour of Lord Rayleigh, one of the pioneers of the subject of acoustics. A few numerical values of characteristic acoustic impedance of fluid and solid media are given in Table 3.1 together with other useful acoustic and physical constants.

Equation (3.35) applies to plane waves propagating in an infinite medium; but when the medium is bounded by a plane surface normal to the direction of propagation, *standing* or *stationary* waves are produced by interference between the direct and reflected waves as shown in Fig. 3.3(a). Particle motion under these conditions has different amplitudes at different points in

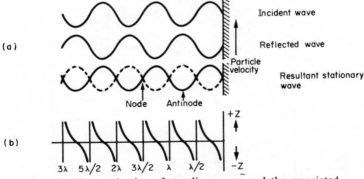

FIG. 3.3. The production of standing waves and the associated specific acoustic impedance. (There is a 90° phase shift in time between the incident and reflected waves so they cannot be added directly).

TABLE 3.1

Material	Density ρ (kg/m³)	Bulk modulus κ (newtons/m³)	Propagation velocity c (m/sec)	Characteristic acoustic impedance ρc (rayls)	Viscosity μ (newton-sec/m²)	Specific heat C_p (joules/kg°C)	Ratio of specific heats γ	Thermal conductivity K (joules/m sec°C)
Fresh-water (20°C)	998	$2 \cdot 18 \times 10^9$	1481	$1 \cdot 48 \times 10^6$	0·001	4192 (10°C)	1·004	0·528
Sea-water (13°C)	1026	$2 \cdot 28 \times 10^9$	1500	$1 \cdot 54 \times 10^6$	0·001	3920 (17°C)	1·01	—
Castor oil (20°C)	950	—	1540	$1 \cdot 45 \times 10^6$	0·96	2130 (20°C)	—	—
Aluminium	2700	$7 \cdot 5 \times 10^{10}$	6300	17×10^6	—	—	—	—
Brass	8500	$13 \cdot 6 \times 10^{10}$	4700	40×10^6	—	—	—	—
Lead	11300	$4 \cdot 2 \times 10^{10}$	2050	$23 \cdot 2 \times 10^6$	—	—	—	—
Nickel	8800	$19 \cdot 0 \times 10^{10}$	5850	$51 \cdot 5 \times 10^6$	—	—	—	—
Steel	7700	$17 \cdot 0 \times 10^{10}$	6100	47×10^6	—	—	—	—
Glass	2300	$3 \cdot 9 \times 10^{10}$	5600	$12 \cdot 9 \times 10^6$	—	—	—	—
Quartz	2650	$3 \cdot 3 \times 10^{10}$	5750	$15 \cdot 3 \times 10^6$	—	—	—	—
Rubber	1000	$0 \cdot 24 \times 10^{10}$	1550	$1 \cdot 55 \times 10^6$	—	—	—	—

space, the stationary points of zero and maximum motion being known as *nodes* and *antinodes* respectively.

In a region where this type of stationary wave is formed, the acoustic pressure is given by eqn. (3.15)

$$p = A \exp [j (\omega t - kx)] + B \exp [j (\omega t + kx)]$$

Following the same lines as the previous case the corresponding particle velocity is found from eqn. (3.20):

$$u = \frac{1}{\rho_0 c} \{A \exp [j (\omega t - kx)] - B \exp [j (\omega t + kx)]\} \quad (3.35)$$

The specific acoustic impedance is therefore

$$Z = \frac{p}{u} = \rho_0 c \left[\frac{A \exp (-jkx) + B \exp (jkx)}{A \exp (-jkx) - B \exp (jkx)} \right] \quad (3.36)$$

If the boundary, positioned at a distance l from the source of plane waves, is perfectly reflecting, the particle velocity u_l at the boundary must be zero, i.e.

$$u_l = \frac{1}{\rho_0 c} [A \exp (-jkl) - B \exp (jkl)] = 0$$

therefore

$$A \exp (-jkl) = B \exp (jkl)$$

Using this result, eqn. (3.36) can be rewritten† as

$$Z = \rho_0 c \left\{ \frac{\exp [jk (l - x)] + \exp [-jk (l - x)]}{\exp [jk (l - x)] - \exp [-jk (l - x)]} \right\}$$
$$= - j\rho_0 c \cot k (l - x) \quad (3.37)$$

This expression shows that the specific acoustic impedance is reactive throughout the space between the source and the reflecting plane. It is zero whenever $\cot k (l - x) = 0$, i.e. whenever

$$l - x = (2n - 1) \lambda/4 \text{ where } n = 1, 2, 3 \text{ etc.}$$

and alternates between $-\infty$ and $+\infty$ every half wavelength in the manner shown in Fig. 3.3b.

† Note that: $\exp (jz) + \exp (-jz) = 2 \cos z$, $\exp (jz) - \exp (-jz) = 2 j \sin z$.

When the boundary is not a perfect reflector, the impedance expression becomes more difficult to derive. (Note that eqns. (3.15) and (3.35) are similar in form to those developed in terms of voltage and current, respectively, for the analysis of high-frequency electrical transmission line phenomena; therefore, plane acoustic waves propagating in a semi-infinite medium can be regarded as analogous to electrical waves in a transmission line and this often proves very useful.)

Consider the case of two media with characteristic acoustic impedances $\rho_0 c_0$, $\rho_T c_T$ respectively, in contact at a distance $x = l$ from the plane wave source which is immersed in the first medium at $x = 0$. The specific acoustic impedance in medium 2, at $x = l$, is given by eqn. (3.36) as

$$Z_T = \frac{p}{u} = Z_0 \left\{ \frac{A \exp(-jkl) + B \exp(jkl)}{A \exp(-jkl) - B \exp(jkl)} \right\}$$

i.e.

$$\frac{B \exp(jkl)}{A \exp(-jkl)} = \frac{Z_T - Z_0}{Z_T + Z_0}$$

therefore

$$\frac{B}{A} = \frac{Z_T - Z_0}{Z_T + Z_0} \exp(-2jkl) \qquad (3.36a)$$

The impedance presented to the plane wave source, Z_S, is also given by eqn. (3.36) in which $x = 0$, i.e.

$$Z_S = Z_0 \frac{A + B}{A - B} = Z_0 \frac{1 + B/A}{1 - B/A} \qquad (3.36b)$$

Combining eqns. (3.36a) and (3.36b) gives

$$Z_S = Z_0 \left[\frac{1 + \dfrac{Z_T - Z_0}{Z_T + Z_0} \exp(-2jkl)}{1 - \dfrac{Z_T - Z_0}{Z_T + Z_0} \exp(-2jkl)} \right]$$

$$= Z_0 \left[\frac{(Z_T + Z_0) \exp(jkl) + (Z_T - Z_0) \exp(-jkl)}{(Z_T + Z_0) \exp(jkl) - (Z_T - Z_0) \exp(-jkl)} \right]$$

$$= Z_0 \left\{ \frac{\begin{aligned} Z_0 \left[\exp (jkl) - \exp (-jkl)\right] \\ + Z_T \left[\exp (jkl) + \exp (-jkl)\right] \end{aligned}}{\begin{aligned} Z_0 \left[\exp (jkl) + \exp (-jkl)\right] \\ + Z_T \left[\exp (jkl) - \exp (-jkl)\right] \end{aligned}} \right\}$$

$$= Z_0 \left(\frac{Z_0 \, 2 \, j \sin kl + Z_T \, 2 \cos kl}{Z_0 \, 2 \cos kl + Z_T \, 2 \, j \sin kl} \right)$$

Dividing the numerator and the denominator by $Z_0 \cos kl$ gives

$$Z_S = \rho_0 c_0 \frac{(\rho_T c_T / \rho_0 c_0) + j \tan kl}{1 + j \, (\rho_T c_T / \rho_0 c_0) \tan kl} \qquad (3.38)$$

One important result revealed by examination of this expression is that when the length of the acoustic medium is equal to one-quarter of a wavelength

$$Z_S = (\rho_0 c_0)^2 / \rho_T c_T \qquad (3.39)$$

The impedance of one acoustic medium at a specified plane normal to the wave direction can therefore be *transformed* by interposing a second medium, $\lambda/4$ in thickness of different characteristic acoustic impedance between the specified plane and the first medium. This second medium forms a matching section which is known as a *quarter-wave transformer*.

6. Spherical Waves in an Infinite Homogeneous Medium

The equation of a spherical wave is obtained from eqn. (3.14) by expressing the Laplacian operator in spherical coordinates[2]

$$\nabla^2 = \frac{\partial^2}{\partial r^2} + \frac{2}{r} \frac{\partial}{\partial r} + \frac{1}{r^2 \sin \theta} \frac{\partial}{\partial \theta} \left(\sin \theta \, \frac{\partial}{\partial \theta} \right) + \frac{1}{r^2 \sin^2 \theta} \frac{\partial^2}{\partial \phi^2} \quad (3.40)$$

using the coordinate transformations $x = r \sin \theta \cos \phi$, $y = r \sin \theta \sin \phi$, $z = r \cos \theta$, shown in Fig. 3.4.

Spherical waves possess spherical symmetry and are functions of radial distance and time but not of the angular coordinates θ and ϕ. Partial derivatives with respect to θ and ϕ are zero, there-

fore, so the final two terms of eqn. (3.40) can be ignored and
eqn. (3.14) becomes

$$\frac{\partial^2 p}{\partial t^2} = c^2 \left(\frac{\partial^2 p}{\partial r^2} + \frac{2}{r} \frac{\partial p}{\partial r} \right)$$

$$= c^2 \left[\frac{1}{r} \cdot \frac{\partial^2}{\partial r^2} (rp) \right] \tag{3.41}$$

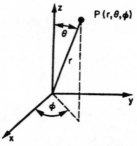

Fig. 3.4. Transformation between cartesian and spherical
coordinates.

The radial coordinate r is independent of time; hence

$$\frac{\partial^2}{\partial t^2} (rp) = r \cdot \frac{\partial^2 p}{\partial t^2}$$

Combining this with eqn. (3.41) leads to the spherical wave
equation

$$\frac{\partial^2}{\partial t^2} (rp) = c^2 \frac{\partial^2}{\partial r^2} (rp) \tag{3.42}$$

which has a general solution

$$p = \frac{1}{r} F_1 (ct - r) + \frac{1}{r} F_2 (ct + r)$$

the two components represent *diverging* and *converging* waves
respectively. This inverse dependence of pressure on distance r is,
of course, in accordance with the well known "inverse-square
law" of energy spreading.

The converging solution is usually ignored as it implies very large acoustic pressures near the origin of the system which invalidates the assumption of low particle amplitude made in its derivation. We assume that the conditions are such that only the diverging wave exists. Once again, the most important solution is one in which the divergent waves perform simple harmonic vibrations represented either in complex form by

$$p = \frac{A}{r} \exp [j(\omega t - kr)] \qquad (3.43a)$$

or in real terms by

$$p = \frac{A}{r} \cos (\omega t - kr) \qquad (3.43b)$$

Many of the relationships between various acoustic parameters derived in connection with plane waves in Section 5 can be applied to spherical waves also, provided a suitable change of coordinate is made.

Thus changing the spatial coordinate to r in eqn. (3.19) gives the radial particle acceleration for a spherical wave

$$a_r = -\frac{1}{\rho_0} \frac{\partial p}{\partial r}$$

from which can be obtained, by integration, the radial particle velocity

$$u_r = -\frac{1}{\rho_0} \int \frac{\partial p}{\partial r} \, \mathrm{d}t$$

Further integration with respect to time gives the radial particle displacement

$$\xi_r = -\frac{1}{\rho_0} \int \int \frac{\partial p}{\partial r} \, \mathrm{d}t \, \mathrm{d}t'$$

For simple harmonic spherical waves these last two expressions become

$$u_r = -\frac{1}{j\omega\rho_0} \cdot \frac{\partial p}{\partial r} \qquad (3.44)$$

and

$$\xi_r = \frac{1}{\omega^2 \rho_0} \cdot \frac{\partial p}{\partial r}$$

respectively.

The evaluation of the coefficient A in eqn. (3.43) presents more of a problem than for the corresponding plane wave case considered earlier. It is usual to relate it to the strength of the source which is defined in the next section.

6.1. *Radiation from a Spherical Source*

The radial acoustic pressure may be regarded as being due to a pulsating spherical source, whose dimensions are small compared with the acoustic wavelength. The radius r_0 of this spherical source varies sinusoidally with time and its surface particle velocity u_s can be represented by

$$u_s = \hat{u} \exp(j\omega t)$$

where $\hat{u}$ is its maximum value.

This type of source is in effect a *point source* and as such is an entirely theoretical concept. However, many real sources, both spherical and plane, provided they are small in comparison with the wavelength can be regarded as point sources.

The radial particle velocity u_r at any point in the medium at a distance r from this type of source is given by eqn. (3.44), using the spatial derivative of eqn. (3.43a), as

$$u_r = \frac{kA}{\omega \rho_0 r} \exp[j(\omega t - kr)] + \frac{A}{j\omega \rho_0 r^2} \exp[j(\omega t - kr)]$$

The rate of flow of medium fluid through a spherical shell of radius r centered on the source is therefore

$$4\pi r^2 u_r = \frac{4\pi r\, kA}{\omega \rho_0} \cdot \exp[j(\omega t - kr)]$$
$$+ \frac{4\pi A}{j\omega \rho_0} \cdot \exp[j(\omega t - kr)] \quad (3.45)$$

The fluid medium must remain in contact with the source at all times, and consequently at the surface of the source the rate of flow of fluid is

$$4\pi r_0^2 \hat{u} \exp (j\omega t)$$

This can be equated to a corresponding expression obtained from eqn. (3.45) in which $r \to r_0$ and $kr \to 0$ since $r = r_0 \ll \lambda$

Thus

$$4\pi r_0^2 \hat{u} \exp (j\omega t) = \frac{4\pi A \exp (j\omega t)}{j\omega\rho_0}$$

from which

$$A = \frac{4\pi r_0^2 \hat{u}}{4\pi} \cdot j\omega\rho_0$$

The product of the surface area and peak particle velocity, $4\pi r_0^2 \hat{u}$, defines the *source strength*. Denoting this quantity by M the coefficient A is given by

$$A = j\omega\rho_0 M/4\pi$$

and the expression for acoustic pressure, eqn. (3.43a), becomes

$$p = \frac{j\omega\rho_0 M}{4\pi r} \cdot \exp [j (\omega t - kr)]$$

which may be expressed in real terms as

$$p = -\frac{\rho_0 c \, kM}{4\pi r} \sin (\omega t - kr)$$

6.2. *The Energy Density and Acoustic Intensity of a Spherical Wave*

The mean energy density of a spherical wave is given by

$$\bar{\mathscr{E}} = \frac{\hat{p}^2}{2\rho_0 c^2} \left(1 + \frac{1}{2k^2 r^2}\right)$$

and when kr is very large (at long ranges), it reduces to the expression derived for a plane wave (eqn. (3.31)). The acoustic intensity of a spherical wave is given by

$$I = \frac{\hat{p}^2}{2\rho_0 c}$$

which is everywhere identical with the corresponding equation for plane waves.

The derivation of these two expressions is far too involved to be included here, and the reader is referred to the work of Kinsler and Frey[3] where it is set out in full.

6.3. *Specific Acoustic Impedance*

An acoustic pressure

$$p = \frac{A}{r} \exp\left[j\left(\omega t - kr\right)\right]$$

gives rise to a radial particle velocity

$$u_r = \frac{A}{\omega \rho_0 r}\left(k + \frac{1}{jr}\right) \exp\left[j\left(\omega t - kr\right)\right]$$

The specific acoustic impedance presented by a fluid medium to a spherical wave as defined by eqn. (3.34) is therefore

$$Z = \frac{p}{u} = \frac{\omega \rho_0}{k + 1/jr} = \frac{j\rho_0 ckr}{1 + j\,kr}$$

This expression can be separated into real and imaginary parts

$$Z = \rho_0 c \cdot \frac{k^2 r^2}{1 + k^2 r^2} + j\,\rho_0 c \cdot \frac{kr}{1 + k^2 r^2}$$

where the first term represents the specific acoustic resistance of the medium and the magnitude of the second term is the specific acoustic reactance.

In the vicinity of a spherical wave source the value of kr is small and the specific acoustic impedance approaches zero. At long ranges when kr is very large, the reactive term approaches zero and the impedance approaches the plane wave value, $\rho_0 c$.

Specific acoustic impedance can be expressed in the alternative form

$$Z = |Z| \cdot \exp(j\theta)$$

where

$$|Z| = \frac{\rho_0 c \, kr}{\sqrt{(1 + k^2 r^2)}}$$

and

$$\theta = \tan^{-1} 1/kr$$

The phase difference is nearly 90° for small values of kr but steadily decreases as kr is increased until the plane wave condition is reached when acoustic pressure and particle velocity are in phase.

It is apparent that

$$\cos \theta = \frac{kr}{\sqrt{(1 + k^2 r^2)}}$$

therefore, the magnitude of the impedance

$$|Z| = \rho_0 c \cos \theta$$

and the relationship between acoustic pressure and particle velocity can be expressed as follows

$$p = \rho_0 \, cu \cos \theta$$

A comparison between certain spherical wave acoustic parameters and their corresponding plane wave values is given in Table 3.2.

7. Attenuation of Acoustic Waves

Except under the special conditions discussed in Chapter 4, Section 7.2, the intensity of initially divergent acoustic waves decreases continuously as they are propagated, the decrease of intensity being referred to as the *propagation loss*. In a few special circumstances, *interference* associated with reflections from boundaries and *diffraction* caused by inhomogeneities in the

TABLE 3.2. PLANE AND SPHERICAL ACOUSTIC WAVE PARAMETERS
(Expressed in real terms)

Parameter	Symbol	Plane wave	Spherical wave
Particle displacement	ξ	$\xi \sin(\omega t - kx)$	$\dfrac{M}{4\pi cr}\cos(\omega t - kr) + \dfrac{M}{4\pi \omega r^2}\sin(\omega t - kr)$
Particle velocity	u	$\omega\xi\cos(\omega t - kx)$	$-\dfrac{kM}{4\pi}\sin(\omega t - kr) + \dfrac{M}{4\pi r^2}\cos(\omega t - kr)$
Particle acceleration	a	$-\omega^2\xi\sin(\omega t - kx)$	$-\dfrac{\omega kM}{4\pi}\cos(\omega t - kr) - \dfrac{\omega M}{4\pi r^2}\sin(\omega t - kr)$
Acoustic pressure	p	$\rho_0 c\tilde{u}\cos(\omega t - kx)$	$-\dfrac{\rho_0 ckM}{4\pi r}\sin(\omega t - kr)$
Propagation velocity	c	$\left(\dfrac{\partial p}{\partial \rho}\right)_{AD} = \left(\dfrac{\gamma\kappa}{\rho_0}\right)^{1/2}$	$\left(\dfrac{\partial p}{\partial \rho}\right)_{AD} = \left(\dfrac{\gamma\kappa}{\rho_0}\right)^{1/2}$
Acoustic intensity	I	$\tfrac{1}{2}p^2/\rho_0 c$	$\tfrac{1}{2}p^2/\rho_0 c$

medium together with *refraction* account for some of this propagation loss but it is mainly attributable to geometrical spreading of the waves and acoustic absorption and scattering by the medium.

The geometrical *spreading loss* associated with spherical waves in an infinite medium is in accordance with the inverse-square law, the intensities I_1 and I_2 at two points P_1 and P_2 in the medium, separated from the spherical wave source by distances r_1 and r_2 respectively, being related as follows

$$I_2 = I_1 (r_1/r_2)^2 \qquad (3.46)$$

If the medium is bounded, as in the case of the ocean, the inverse square law does not always apply; it is then usual to modify eqn. (3.46) to

$$I_2 = I_1 (r_1/r_2)^n$$

where n is a numerical exponent with a non-integral value of less than two. During long-range, shallow-water propagation the spreading becomes cylindrical and the exponent n is unity.

The combined effects of scattering and absorption within the medium produce the second basic component of propagation loss, the *attenuation loss*. Usually the fractional reduction of intensity per unit distance is constant; therefore, if the spreading loss is included, the intensities at points P_1 and P_2 are related by the expression

$$I_2 = I_1 (r_1/r_2)^n \exp [- 2a (r_2 - r_1)] \qquad (3.47)$$

where a is the absorption coefficient measured in nepers/m.

Equation (3.47) is often written in the form

$$10 \log_{10} I_1 - 10 \log_{10} I_2 = 10n \log_{10} r_2/r_1 + 8 \cdot 7a(r_2 - r_1) \quad (3.48)$$

where the left-hand side represents the decrease in intensity as the wave is propagated from P_1 to P_2 and is therefore the propagation loss N measured in decibels. It is customary to replace $8 \cdot 7a$ by a where a is the absorption coefficient measured in dB/m. Since the overall propagation loss can strictly be defined only between two finite ranges r_1 and r_2 one cannot strictly refer to the loss from the source to a point at distance r_2. This difficulty is overcome by taking an *index point* at a distance 1 m from the effective centre of the sound source and saying that the propagation loss at a range r is that corresponding to the loss between the index point and r. The significance of this is made clear in Section 8. Therefore eqn. (3.48) can be written in the simplified form

$$N \simeq 10n \log_{10} r + ar \text{ dB} \qquad (3.49)$$

Propagation loss measurements made at sea are often far larger than those predicted by this equation due to the effects of interference, diffraction, reflection and refraction. Although it is possible to compute the effects of these factors individually under idealized conditions, the characteristics of the ocean are so changeable that it is better to combine contributions from these factors under a single term H known as the *transmission anomaly* and modify eqn. (3.49) according to

$$N \simeq 10n \log_{10} r + ar + H \text{ dB} \qquad (3.49a)$$

At this point it is necessary to investigate in detail the various causes of absorption in fluid media, and in particular in the sea, and to derive expressions for its magnitude expressed in terms of the absorption coefficient and to determine the dependence of this coefficient upon the frequency of the propagating wave.

As an acoustic wave propagates, all of its energy is ultimately converted into heat by the non-ideal nature of the medium. In fluid media, the effects of *viscosity*, *thermal conduction*, *intramolecular* and *relaxation* phenomena all take part in this heat-conversion process.

7.1. *Attenuation due to Viscosity*

As a simple introduction to the involved processes of sound absorption, viscous attenuation of plane waves in an infinite medium will be considered in detail. During the propagation of such a wave, different regions of the medium move relative to one another. Viscous forces, which partially oppose the stresses set up by the acoustic pressure, are experienced and energy is extracted from the acoustic wave to overcome them. This energy is not merely stored as with elastic forces, it is actually converted and lost to the acoustic wave. In the presence of these viscous forces, the acoustic pressure is not constant in all directions and Rayleigh[4] showed that p in eqn. (3.12) should be replaced by

$$p - \frac{4}{3} \mu \frac{\partial}{\partial t} \left(\frac{\partial \xi}{\partial x} \right)$$

for plane waves propagating in the x direction of Fig. 3.2. In this expression, μ is the longitudinal coefficient of shear viscosity defined as

$$\frac{\text{the shear stress in a plane}}{\text{particle velocity gradient normal to the plane}}$$

measured in newton sec/m^2.

Equation (3.12) therefore becomes

$$\rho_0 \frac{\partial u}{\partial t} = \rho_0 \frac{\partial^2 \xi}{\partial t^2} = - \frac{\partial p}{\partial x} + \frac{\partial}{\partial x} \left[\frac{4}{3} \mu \frac{\partial}{\partial t} \left(\frac{\partial \xi}{\partial x} \right) \right] \qquad (3.50)$$

The first derivative of eqn. (3.4) with respect to x can be combined with eqn. (3.25) to give

$$- \frac{\partial p}{\partial x} = \kappa \frac{\partial^2 \xi}{\partial x^2}$$

Using this expression eqn. (3.50) becomes

$$\rho_0 \frac{\partial^2 \xi}{\partial t^2} = \kappa \frac{\partial^2 \xi}{\partial x^2} + \frac{4}{3} \mu \frac{\partial}{\partial t} \left(\frac{\partial^2 \xi}{\partial x^2} \right) \qquad (3.51)$$

This is the *attenuated plane wave equation* in a viscous medium expressed in terms of particle displacement.

Assuming the usual simple harmonic form of solution

$$\xi = d \exp [j\omega (t - x/c_1)] \qquad (3.52)$$

where d is the peak displacement and c_1 is the velocity of propagation in the viscous medium,

$$\frac{\partial^2 \xi}{\partial x^2} = \frac{- \omega^2}{c_1^2} \xi, \quad \frac{\partial^2 \xi}{\partial t^2} = - \omega^2 \xi \quad \text{and} \quad \frac{\partial}{\partial t} \left(\frac{\partial^2 \xi}{\partial x^2} \right) = - j \frac{\omega^3}{c_1^2} \xi$$

Equation (3.51) therefore reduces to

$$1 = \frac{\kappa}{\rho_0 c_1^2} + \frac{jA}{c_1^2}$$

where

$$A = \frac{4}{3} \frac{\mu\omega}{\rho_0}$$

From this can be obtained the propagation velocity c_1 in the viscous medium in terms of the velocity in an ideal medium, $c = \sqrt{(\kappa/\rho_0)}$:

$$\frac{1}{c_1^2} = \frac{1}{c^2} (1 - jA/c^2)^{-1}$$

which is complex. Usually $A \ll 1$ and the propagation velocity is given by

$$\frac{1}{c_1} = \frac{1}{c} \left(1 - j \frac{A}{2c^2}\right) \tag{3.53}$$

Substitution of this expression into eqn. (3.52) gives

$$\xi = d \exp [j\omega (t - x/c + j \, Ax/2c^3)]$$
$$= d \exp (- a_v x) \exp [j (\omega t - \beta x)]$$

which is the equation of an attenuated plane wave propagating with velocity $c = \omega/\beta$. In this expression a_v is the viscous absorption coefficient and is given by

$$a_v = \tfrac{2}{3} \, v\omega^2/c^3 \tag{3.54}$$

where v is the kinematic viscosity measured in m^2/sec, and replaces μ/ρ_0. It should be noted that a_v is dependent upon the square of the frequency of the acoustic wave.

7.2. *Attenuation due to Thermal Conduction*

The pressure variations occurring during acoustic propagation are assumed to be adiabatic and consequently they are accompanied by changes of temperature. A thermal gradient is produced and heat flows from regions of compression to regions of rarefaction. This heat flow tends to equalize the pressure differences in the medium and the wave amplitude is decreased accordingly as it propagates. The simplified attenuated plane wave equation in a thermally conducting medium is given by Vigoureux[5] as

$$\frac{\partial^2 \xi}{\partial t^2} = c^2 \frac{\partial^2 \xi}{\partial x^2} + \frac{K}{\rho_0 C_v}\frac{\gamma - 1}{\gamma}\frac{\partial}{\partial t}\left(\frac{\partial^2 \xi}{\partial x^2}\right)$$

where K is the thermal conductivity of the medium measured in joules/m sec°C. This expression is derived as before by replacing the acoustic pressure in eqn. (3.12) by its value after taking account of thermal conductivity and is of the same form as eqn. (3.51). If a sinusoidal solution is assumed once more, the absorption coefficient in a thermally conducting medium can be written by inspection, as

$$a_{tc} = \frac{K}{2\rho_0 C_v}\frac{\gamma - 1}{\gamma}\frac{\omega^2}{c^3} \qquad (3.55)$$

thus revealing the same frequency dependence as viscous attenuation.

Using the physical constants listed in Table 3.1, eqns. (3.54) and (3.55) give the absorption coefficients due to viscosity and thermal conduction in fresh-water as $a_v = 8 \cdot 1 \times 10^{-15} f^2$ and $a_{tc} = 3 \times 10^{-18} f^2$ respectively. Thermal conduction as an attenuating mechanism in water can therefore be neglected. The measured value of a in fresh-water, however, is $24 \times 10^{-15} f^2$ and some explanation for this discrepancy which is known as *excess absorption* must be sought.

7.3. *Excess Absorption due to Relaxation*

One such explanation was offered by Hall[6] using a theory of structural relaxation. This theory assumes that water is capable of existing in two energy states, the state of lower energy being the normal one, the high energy state being one in which the molecules are more closely packed. Under static conditions most of the molecules are in the normal state but the presence of a compressional acoustic pressure wave causes some of the molecules to pass into the more closely packed state, energy being extracted from the acoustic wave to do so. When the molecules revert to the normal state the energy is returned to the acoustic wave.

The finite time known as the *relaxation time* τ is required for molecules to pass from the lower state into the upper state and back again, however. At low acoustic frequencies when the period of the wave is long compared with the relaxation time, the reconverted energy is in phase with the energy extracted from the wave and no attenuation is experienced. At high frequencies no time is allowed for the molecules to change their states so once again no attenuation occurs.

It is to be expected, therefore, that when the period of the wave is comparable with the relaxation time, energy stored in the higher state as a result of a compressional cycle will reappear during the subsequent rarefaction cycle thereby tending to equalize the acoustic pressure difference and causing high attenuation of the wave.

The frequency dependence of the attenuation coefficient describing such a relaxation process is of the characteristic form derived by Vigoureux[5]

$$a = 2a_m \frac{f^2 f_r}{f^2 + f_r^2} \tag{3.56}$$

where a_m is the maximum value of a occurring at the relaxation frequency, $f_r = 1/2\pi\tau$.

Direct measurements by Liebermann[7] indicate that this structural relaxation increases the effective kinematic viscosity of water by a factor of 3. Using this modified value, eqn. (3.54) gives a value of a_v which is found to be in satisfactory agreement with the measured value.

The measured attenuation of acoustic waves in fresh-water and salt-water is shown in curves A and B respectively of Fig. 3.5. Below 500 kc/s, the pronounced discrepancy between these two curves suggests yet another relaxation process. This can be attributed to the presence of dissolved salts, in particular magnesium sulphate, the process being known as a *chemical relaxation*. In water, magnesium sulphate dissociates into positive and negative ions, dissociation and recombination being in a state of equilibrium. However, the passage of a compressional wave causes excess recombination of these ions. The time delays of this process and the subsequent redissociation lead to relaxational dissipation

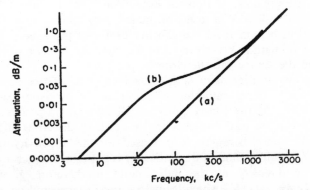

FIG. 3.5. Attenuation coefficients at ultrasonic frequencies, (a) in fresh-water at 5°C and (b) in sea-water at 5°C.

of acoustic energy characterized as before by eqn. (3.56). Attenuation loss in sea-water is due solely to viscous attenuation and chemical relaxation. The attenuation coefficient a in eqn. (3.49) must therefore be of the form

$$a = Af^2 + \frac{Bf_rf^2}{f_r^2+f^2} \text{ dB/m} \qquad (3.57)$$

where it must be remembered $a = 8\cdot7\,\alpha$. In this expression, the relaxation frequency of the dissociation of magnesium sulphate is dependent upon temperature, increasing from 60 kc/s at 5°C to 100 kc/s at 15°C. The value of A is also temperature dependent, below the frequency of 200 kc/s; however, these temperature variations have little effect upon the total attenuation in sea-water. If f and f_r are expressed in kc/s, measurements in the ocean indicate that the constants in eqn. (3.57) which fit the empirical data most exactly are $A = 3 \cdot 2 \times 10^{-7}$ and $B = 6 \times 10^{-4}$.

8. Factors Influencing the Useful Range of Underwater Acoustic Systems

In the previous section an expression was obtained for the propagation loss N in terms of the intensity relative to an index point 1 m distant from the effective centre of the source. If the source strength is known the acoustic intensity I_R at any distance r from the source can therefore be predicted.

The acoustic intensity I_0 at a distance of 1 m from an omnidirectional source radiating a power of W acoustic watts is $I_0 = W/4\pi$. If the source possesses directional properties more energy can be transmitted along a specified direction and the effective source strength is increased by the directivity factor (D.F.). (A full account of directionality is given in Chapter 6.) The acoustic intensity at the index point is therefore given by $I_0 = W/4\pi \times$ (D.F.) which leads to a *source intensity* level of

$$I_S = 10 \log_{10} I_0 = - 11 + 10 \log_{10} W + \text{D.I.S.} \quad (3.58)$$

in decibels relative to 1 W/m² where D.I.S., the directivity index, equals $10 \log_{10}$ D.F. The acoustic intensity level I_R is, at a distance r from the source, therefore given by

$$I_R = I_S - N$$

decibels relative to 1 W/m². This equation can be used to determine the theoretical maximum separation r between the transmitter and the receiver of a communication or telemetry system if the omnidirectional noise experienced at the receiver, is taken into account. The intensity I_n of this noise depends upon the centre frequency and bandwidth of the receiver[3]. This noise is reduced by the directionality D.I. of the receiver, however; thus the effective noise intensity is given by

$$I_N = 10 \log_{10} I_n - \text{D.I.R.}$$

decibels relative to 1 W/m^2. For reliable communication a *recognition differential* D, measured in decibels, is required to ensure that the signal is not masked by noise; therefore at the receiver

$$I_R \geqslant I_N + D$$

Expressed in full and using equation (3.49a), this gives

$$10n \log_{10} r + ar + H \leqslant -11 + 10 \log_{10} W$$
$$+ \text{D.I.S.} + \text{D.I.R.} - 10 \log_{10} I_N - D$$

from which the maximum separation r can be found.

Three additional factors affect the maximum range of underwater echo-ranging equipment;

(a) The reflecting properties of the target object, characterized by its *target strength,*
(b) The return path over which the signal is, once again, subjected to spreading and absorption losses, and
(c) The increase in noise masking the received signal due to reverberation.

The target strength T of a reflecting body is defined by the expression

$$T = 10 \log_{10} I_s/I_I$$

where I_s and I_I represent the reflected (or scattered) and incident intensities respectively. For complex targets, target strength is difficult to calculate or measure, but for a smooth spherical rigid target a simple, if approximate, method can be used which yields a result in agreement with measured values. There is only a single echo and if it is assumed that the diameter of the sphere is large compared with the acoustic wavelength and that the sphere re-radiates equally in all directions, the calculation proceeds as follows.

The source strength is defined as the sound intensity at 1 m from the source. If the sound field at the target has intensity I and the sphere has radius a_1 metres, then the sound power incident upon the target is $\pi a_1^2 I$. This power is reradiated equally in all directions so that at 1 m from the centre of the sphere the sound intensity is $\pi a_1^2 I/4\pi = a_1^2 I/4$. This then is the actual source strength of the target. The source strength relative to the intensity of the incident field is therefore $a_1^2/4$. This is the target strength; expressing it according to our definition it becomes

$$T = 20 \log_{10} a_1/2 \text{ dB}$$

and a unit target, having a target strength of 0 dB, will be a sphere of 2 m radius. The target strength of an object with irregular shape depends not only on its size but also on its orientation. For each particular orientation it may be assumed to present a different *effective scattering cross section* σ measured in m². The corresponding target strength is then given by

$$T = 10 \log_{10} (\sigma/4\pi) \text{ dB}$$

Reverberation is the name given to the combined signals reflected by objects in the sea other than the target; these objects may take the form of bubbles, suspended particles, inhomogeneities in the water, fish, the sea-surface or the sea-bed. These unwanted signals arrive at the receiver at the same time as the signal from the target and cause additional masking of the required signal.

Taking account of factors (a) and (b) the echo intensity level I_E at the receiver of an echo-ranging equipment is given by

$$I_E = I_S - N + T - N = I_S + T - 2N$$

If the reverberation intensity level R is defined in the usual manner, i.e.

$$R = 10 \log_{10} R_I$$

where R_I is the reverberation intensity measured in W/m², the maximum range of an echo-ranger is in accordance with the equation

$$I_E \geqslant 10 \log_{10} (I_n + R_I)$$

which can be written, in full, in the form

$$20n \log_{10} r + 2ar + 2H \leqslant -11 + 10 \log_{10} W + \text{D.I.S.}$$

$$+ \text{D.I.R.} + 10 \log_{10} \frac{\sigma}{4\pi} - 10 \log_{10} (I_n + R_I) - D$$

from which the value of r can be found.

9. Non-linear Effects due to High-amplitude Waves

We have seen in Section 3 that the elementary theory of acoustic waves requires the assumption that the bulk modulus κ is independent of the acoustic pressure, and similarly in Section 4 we have assumed that the density ρ is constant. This means that the theory developed so far is strictly only valid for waves of very low amplitude, since for waves of finite amplitude both κ and ρ vary (if only slightly) according to the acoustic pressure. Thus instead of the velocity of propagation, c, being a constant for given conditions, it actually varies according to the waveform and amplitude of the acoustic wave. If at any point in the medium the acoustic pressure is p, then eqn. (3.6) can be replaced by

$$c^2 = (\kappa_0/\rho_0)(1 + a_1 p + a_2 p^2 + \ldots) \tag{3.59}$$

where a_1, a_2, etc., are constants determined by the parameters of the medium. The wave equation (3.14) for a plane wave then becomes

$$\frac{\partial^2 p}{\partial t^2} = \frac{\kappa_0}{\rho_0}(1 + a_1 p + a_2 p^2 + \ldots) \nabla^2 p \tag{3.60}$$

The physical interpretation of this is clearly that if a sinusoidal pressure–time waveform is imposed at the spatial origin (i.e. $x = 0$), then the spatial (pressure–distance) waveform is not sinusoidal, and therefore the pressure–time waveform at any point other than $x = 0$ is not sinusoidal, but contains harmonics.

If two or more sinusoidal waveforms are impressed on the system, clearly sum and difference frequencies are also generated. Practical exploitation of these effects seems possible, and indeed promising.[8, 9, 10, 11]

References

1. TUCKER, D. G., *Elementary Electrical Network Theory*, Pergamon, London (1964).
2. HAGUE, B., *An Introduction to Vector Analysis for Physicists and Engineers*, Methuen's Monographs on Physical Subjects, London (1959).
3. KINSLER, L. E., and FREY, A. R., *Fundamentals of Acoustics*, 2nd Edition, Wiley, New York (1962).

4. RAYLEIGH, LORD, *Theory of Sound*, Dover, New York (1945).

5. VIGOUREUX, P., *Ultrasonics*, Chapman and Hall, London (1950).

6. HALL, L., The Origin of Ultrasonic Absorption in Water, *Phys. Rev.* **73**, 775 (1948).

7. LIEBERMANN, L. N., The Second Viscosity of Liquids, *Phys. Rev.* **75**, 1415 (1949).

8. TUCKER, D. G., The Exploitation of Non-linearity in Underwater Acoustics, *J. Sound Vib.* **2**, 429 (1965).

9. BERKTAY, H. O., Possible Exploitation of Non-linear Acoustics in Underwater Transmitting Applications, *J. Sound Vib.* **2**, 435 (1965).

10. BERKTAY, H. O., Parametric Amplification by the Use of Acoustic Non-linearities and Some Possible Applications, *J. Sound Vib.* **2**, 462 (1965).

11. WELSBY, V. G., KULJIS, M., and DUNN, D. J., Non-linear Effects in a Focused Underwater Standing-wave Acoustic System, *J. Sound Vib.* **2**, 471 (1965).

CHAPTER 4

Acoustic Propagation in the Sea

1. Introduction

In the previous chapter, the propagation of acoustic energy through an ideal, homogeneous, unbounded medium was investigated. In this chapter, it is proposed to consider the propagation of sound through a *real* medium, the sea, which is both grossly inhomogeneous and bounded by its surface and bed. The simple wave equation (3.14) developed earlier, strictly applies only to waves in an infinite homogeneous medium, but provided the dimensions of the boundaries are large compared with the acoustic wavelength and provided the acoustic properties of the medium vary in a regular manner spatially, i.e. have no random characteristics, results obtained by its use are still valid.

In this chapter, certain basic general relationships governing the reflection and transmission of acoustic energy at a boundary between two dissimilar acoustic media will first be derived. Relationships concerning the propagation of sound through a medium with a non-uniform or spatially variable velocity of propagation will also be discussed.

The particular problem of propagating energy through the sea can then be investigated by making use of these various relationships. The problem will be reduced primarily to its simplest terms, then additional factors will be introduced until a picture of the very complex nature of sound propagation in the sea has been built up.

In the sea, we are usually concerned with sound propagation only over paths which are very long compared with the acoustic wavelength, and thus the waves are effectively plane waves. The discussion throughout will therefore be restricted to plane waves and plane boundaries. For a treatment of the reflection and transmission of spherical waves, the reader is referred to such works as Brekhovskikh.[1]

Sound waves impinging upon a plane boundary between two acoustic media undergo reflection and refraction in the same manner as light waves impinging upon an air–glass interface. Simple experiments show that sound waves obey the same simple basic laws, namely:

(a) the angle of incidence is equal to the angle of reflection, and

(b) the incident ray, the reflected ray and the normal to the point of incidence all lie in the same plane.

These two laws and the acoustic equivalent of Snells' Law (see Section 2) enable acoustic reflection and refraction problems to be solved qualitatively. If the problems are to be solved on an analytical basis, however, the following procedure must be adopted.

2. Transmission between Two Fluid Media

Acoustic pressure is constant over any particular wavefront propagating in a loss-less medium and is represented by the expression

$$p = \hat{p} \exp [j(\omega t - kd)]$$

where $d = x \cos \theta + y \sin \theta$ is the distance of the wavefront WW' from the origin of the system along the direction of propagation shown in Fig. 4.1.

If the boundary lies along the y axis as shown in Fig. 4.2 then the equation for the pressure in the incident wave travelling through medium 1 making an angle θ_1 with the positive x-axis is

$$p_i = \hat{p} \exp [j(\omega t - k_1 x \cos \theta_1 - k_1 y \sin \theta_1)] \qquad (4.1)$$

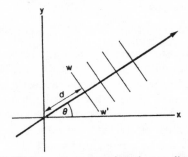

FIG. 4.1. Plane waves propagating along a direction θ.

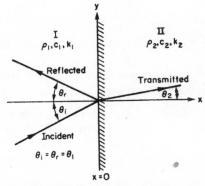

FIG. 4.2. Plane waves incident obliquely upon a plane boundary situated at $x = 0$.

The angle of reflection is equal to the angle of incidence; consequently, the angle between the reflected ray and the positive x-axis is $(180° - \theta_1)$ and the reflected ray is expressed by

$$p_r = V\hat{p} \exp\left[j(\omega t + k_1 x \cos\theta_1 - k_1 y \sin\theta_1)\right] \quad (4.2)$$

The coefficient V represents the fraction of the pressure amplitude reflected from the boundary and is termed the *reflection coefficient*.

The transmitted wave is refracted at the boundary and makes an angle θ_2 with the normal to the boundary in medium 2 and may be represented by

$$p_t = W\hat{p} \exp [\mathrm{j}(\omega t - k_2 x \cos \theta_2 - k_2 y \sin \theta_2)] \qquad (4.3)$$

where W is the *transmission coefficient* and represents the fraction of the pressure amplitude transmitted into the second medium.

At the boundary, when $x = 0$, the acoustic pressure on both sides must be equal, since otherwise the boundary would experience a net force upon it. Also the normal component of the particle velocity must be the same on both sides of the boundary if the two media are to remain in contact.

These two conditions may be expressed mathematically as

$$p = p_i + p_r = p_t \qquad (4.4)$$

and

$$u_i \cos \theta_1 + u_r \cos (180° - \theta_1) = u_t \cos \theta_2 \qquad (4.5)$$

and used to solve eqns. (4.1), (4.2) and (4.3) for the coefficients V and W.

Applying eqn. (4.4) to these equations at $x = 0$ and eliminating the common frequency term $\exp (\mathrm{j}\omega t)$ gives

$$\hat{p} \exp (-\mathrm{j}\, k_1 y \sin \theta_1) + V \hat{p} \exp (-\mathrm{j}\, k_1 y \sin \theta_1)$$
$$= W\hat{p} \exp (-\mathrm{j}\, k_2 y \sin \theta_2) \qquad (4.6)$$

thus

$$1 + V = W \exp [-\mathrm{j}y\, (k_2 \sin \theta_2 - k_1 \sin \theta_1)] \qquad (4.7)$$

The left-hand side of this equation is independent of y and, consequently, the right-hand side must be independent of y also. This can only be true if

$$k_2 \sin \theta_2 = k_1 \sin \theta_1$$

which, using eqn. (3.17), leads directly to Snell's Law relating the angles of incidence and refraction:

$$\frac{\sin \theta_1}{\sin \theta_2} = \frac{c_1}{c_2} = \eta \qquad (4.8)$$

where η is the refractive index between the two media. Applying Snell's Law to eqn. (4.7) gives

$$1 + V = W \qquad (4.9)$$

Using eqn. (3.24), eqn. (4.5) may be expressed in terms of the maximum pressure, thus

$$\frac{\hat{p}}{\rho_1 c_1} \cos \theta_1 - \frac{V\hat{p}}{\rho_1 c_1} \cos \theta_1 = \frac{W\hat{p}}{\rho_2 c_2} \cos \theta_2 \qquad (4.10)$$

i.e.

$$\rho_2 c_2 \cos \theta_1 (1 - V) = \rho_1 c_1 \, W \cos \theta_2 \qquad (4.11)$$

Eliminating W between eqns. (4.9) and (4.11) gives the reflection coefficient

$$V = \frac{\rho_2 c_2 \cos \theta_1 - \rho_1 c_1 \cos \theta_2}{\rho_2 c_2 \cos \theta_1 + \rho_1 c_1 \cos \theta_2} \qquad (4.12)$$

This reflection coefficient applies to particle amplitude, particle velocity, particle acceleration and acoustic pressure; but when incident and reflected intensities are to be compared the *acoustic power reflection coefficient* a_r must be used.

It is given by

$$a_r = \frac{I_r}{I_i} = \frac{(V\hat{p})^2}{2 \, \rho_1 c_1} \Big/ \frac{\hat{p}^2}{2 \, \rho_1 c_1} = V^2$$

$$= \left(\frac{\rho_2 c_2 \cos \theta_1 - \rho_1 c_1 \cos \theta_2}{\rho_2 c_2 \cos \theta_1 + \rho_1 c_1 \cos \theta_2} \right)^2 \qquad (4.13)$$

Eliminating V between eqns. (4.9) and (4.11) gives the transmission coefficient

$$W = \frac{2 \, \rho_2 c_2 \cos \theta_1}{\rho_2 c_2 \cos \theta_1 + \rho_1 c_1 \cos \theta_2} \qquad (4.14)$$

The ratio of transmitted intensity to incident intensity gives the *intensity transmission coefficient* a_t, thus:

$$a_t = \frac{I_t}{I_i} = \frac{(W\hat{p})^2/2 \, \rho_2 c_2}{\hat{p}^2/2 \, \rho_1 c_1} = \frac{4 \, \rho_1 c_1 \, \rho_2 c_2 \cos^2 \theta_1}{(\rho_2 c_2 \cos \theta_1 + \rho_1 c_1 \cos \theta_2)^2} \qquad (4.15)$$

An acoustic beam, however, is widened or narrowed as it passes from one medium to another depending upon whether it is refracted away from or towards the normal to the boundary and the method of obtaining eqn. (4.15) does not take account of this.

An alternative coefficient, the *acoustic power transmission coefficient* a'_t equal to the difference between incident and reflected intensities

$$a'_t = 1 - a_r \qquad (4.16)$$

is usually defined to overcome this; expressed in terms of the medium parameters and the angles of incidence and refraction

$$a'_t = \frac{4 \, \rho_1 c_1 \, \rho_2 c_2 \, \cos \theta_1 \, \cos \theta_2}{(\rho_2 c_2 \, \cos \theta_1 + \rho_1 c_1 \, \cos \theta_2)^2} \qquad (4.17)$$

In all of the above results it is possible to express the angle of refraction θ_2 in terms of the angle of incidence θ_1. This complicates the expressions, however, and will only be used here to obtain specific results.

The expressions of eqns. (4.12), (4.13), (4.14), (4.15) and (4.17) can be simplified by replacing $\rho_1 c_1$ and $\rho_2 c_2$ by R_1 and R_2, the characteristic acoustic impedances (which are strictly pure resistances), and dividing throughout by $\cos^2 \theta_1 \cos^2 \theta_2$; thus the acoustic power reflection and transmission coefficients become

$$a_r = \left(\frac{\dfrac{R_2}{\cos \theta_2} - \dfrac{R_1}{\cos \theta_1}}{\dfrac{R_2}{\cos \theta_2} + \dfrac{R_1}{\cos \theta_1}} \right)^2 \qquad (4.18)$$

and

$$a'_t = \frac{4 \dfrac{R_1}{\cos \theta_1} \dfrac{R_2}{\cos \theta_2}}{\left(\dfrac{R_2}{\cos \theta_2} + \dfrac{R_1}{\cos \theta_1} \right)^2} \qquad (4.19)$$

which can be expressed in the general form

$$a_r = \left(\frac{Z_2 - Z_1}{Z_2 + Z_1} \right)^2 \qquad (4.20)$$

and

$$a'_t = \frac{4 \, Z_1 Z_2}{(Z_1 + Z_2)^2} \qquad (4.21)$$

by replacing $R_1/\cos\theta_1$ and $R_2/\cos\theta_2$ by Z_1 and Z_2 respectively.

The results obtained so far, being for acoustic transmission at oblique incidence, are perfectly general except for the restriction to plane waves, so can be used to investigate certain special cases of acoustic transmission between two liquid media.

2.1. *Transmission between Two Fluid Media, Normal Incidence*

When the incident acoustic pressure wave is normal to the boundary between two media, $\theta_1 = \theta_2 = 0$ and the expressions for the acoustic power reflection and transmission coefficients reduce to

$$a_r = \left(\frac{R_2 - R_1}{R_2 + R_1}\right)^2 \tag{4.22}$$

and

$$a'_t = \frac{4\,R_1 R_2}{(R_2 + R_1)^2} = a_t \tag{4.23}$$

The pressure amplitude reflection coefficient

$$V = \frac{R_2 - R_1}{R_2 + R_1} \tag{4.24}$$

and its sign depends upon the relative magnitudes of R_1 and R_2. When $R_2 \gg R_1$ the reflection coefficient approximates to $+1$ and most of the acoustic energy will be reflected from the boundary without a change in the phase of the acoustic pressure; such a boundary is said to be *rigid*. The phase of the particle displacement, velocity and acceleration will be reversed, however, in accordance with the relationship $p = -\rho_0 c u$ which applies to a reflected wave.

When $R_1 \gg R_2$ the reflection coefficient approximates to -1; once again, most of the acoustic energy will be reflected but the phase of the acoustic pressure will be reversed and the boundary is said to be a *pressure release* or *soft* boundary. The phase of the reflected particle displacement, velocity and acceleration, being opposite to that of the acoustic pressure, will be unchanged. Between these two limits, partial reflection will occur but the phase shift of the acoustic pressure wave will only be zero or π rads, in accordance with eqn. (4.24).

2.2. *The Angle of Intromission*

Examination of eqn. (4.12) shows that the reflection coefficient is zero whenever

$$\rho_2 c_2 \cos \theta_1 - \rho_1 c_1 \cos \theta_2 = 0 \qquad (4.25)$$

Using the well known trigonometrical relationships

$$\cos \theta_1 = \sqrt{(1 - \sin^2 \theta_1)} \qquad (4.26)$$

$$\cos \theta_2 = \sqrt{(1 - \sin^2 \theta_2)}$$

and Snell's Law

$$\sin \theta_2 = (c_2/c_1) \sin \theta_1$$

eqn. (4.25) can be rewritten as

$$\rho_2 c_2 \sqrt{(1 - \sin^2 \theta_1)} - \rho_1 c_1 \sqrt{\left(1 - \frac{c_2^2}{c_1^2} \sin \theta_1^2\right)} = 0$$

From this equation, an expression for the angle of incidence in the first medium at which acoustic transmission between the two media is complete can be obtained, thus

$$\sin^2 \theta_1 = \frac{\rho_2^2 c_2^2 - \rho_1^2 c_1^2}{c_2^2 (\rho_2^2 - \rho_1^2)}$$

$$= \frac{(\rho_2/\rho_1)^2 - (c_1/c_2)^2}{(\rho_2/\rho_1)^2 - 1} \qquad (4.27)$$

This particular angle of incidence is known as the angle of intromission and is analogous to the *Brewster angle* of electromagnetic theory.[2] For this angle to be real, however, certain conditions must be satisfied; when $\rho_2/\rho_1 > 1$ then $1 < (c_1/c_2) < (\rho_2/\rho_1)$ and when $\rho_2/\rho_1 < 1$ then $1 > (c_1/c_2) > (\rho_2/\rho_1)$.

2.3. *Total Internal Reflection*

When the velocity of propagation is greater in the second medium than in the first the angle of refraction is greater than the angle of incidence. If the angle of incidence is increased from zero, some particular angle is reached, known as the *critical angle*, at which the refracted wave becomes coincident with the boundary and the angle of refraction is $\pi/2$ rad. This critical angle is obtained from Snell's Law by substituting $\theta_2 = \pi/2$ into eqn. (4.8) thus

$$\theta_{\text{crit}} = \sin^{-1}(c_1/c_2) \qquad (4.28)$$

If the angle of incidence is greater than this critical value, the refracted wave disappears, the angle of refraction becomes imaginary and *total internal reflection* occurs. Under these conditions the equation for the reflection coefficient, eqn. (4.12), which can be expressed in the form

$$V = \frac{\rho_2 c_2 \cos \theta_1 - \rho_1 c_1 \sqrt{[1 - (c_2/c_1)^2 \sin^2 \theta_1]}}{\rho_2 c_2 \cos \theta_1 + \rho_1 c_1 \sqrt{[1 - (c_2/c_1)^2 \sin^2 \theta_1]}} \qquad (4.29)$$

using eqns. (4.12) and (4.26), becomes

$$V = \frac{\rho_2 c_2 \cos \theta_1 - j\,\rho_1 c_1 \sqrt{[\sin^2 \theta_1 (c_2/c_1)^2 - 1]}}{\rho_2 c_2 \cos \theta_1 + j\,\rho_1 c_1 \sqrt{[\sin^2 \theta_1 (c_2/c_1)^2 - 1]}} \qquad (4.30)$$

which can be written in the simplified form

$$V = \frac{b \cos \theta_1 - j\,\sqrt{(\sin^2 \theta_1 - \eta^2)}}{b \cos \theta_1 + j\,\sqrt{(\sin^2 \theta_1 - \eta^2)}} \qquad (4.30a)$$

where $b = \rho_2/\rho_1$ and $\eta = c_1/c_2$

The modulus of this expression is unity, and consequently all the incident acoustic energy is reflected but the phase of the reflected pressure wave is shifted relative to the phase of the incident pressure wave by an angle ϕ given by the argument of eqn. (4.30a).

$$\phi = -2 \tan^{-1} \frac{\sqrt{(\sin^2 \theta_1 - \eta^2)}}{b \cos \theta_1} \qquad (4.31)$$

2.4. *Internal Reflection when Second Medium has Absorption*

Total internal reflection never takes place, in practice, at any angle of incidence owing to the effect of absorption in the second medium which results in the production of inhomogeneous waves along the boundary.

In Chapter 3, Section 7.1, it was shown that the propagation velocity in an absorbing medium is complex; therefore the acoustic wave number must be complex also and of the form $k_2 = k - ja'$ where k is related to the phase velocity by the equation $k = \omega/c$ and a' is the volume attenuation coefficient. The refractive index between the two media is therefore,

$$\eta = \frac{c_1}{c_2} = \frac{k_2}{k_1} = \frac{k}{k_1} (1 + ja) = \eta_0(1 + ja) \qquad (4.32)$$

where $a = -a'/k$; thus η is also complex.

In most applications it happens that a'/k is sufficiently small to justify the approximation

$$\eta^2 = \eta_0^2 (1 + 2 ja) \qquad (4.33)$$

so the term $\sqrt{(\sin^2 \theta_1 - \eta^2)}$ in eqn. (4.30a) can be written as $\sqrt{(\sin^2 \theta_1 - \eta_0^2 - 2 j\eta_0^2 a)}$ which is of the form $\sqrt{(A - jB)}$ where $A = \sin^2 \theta_1 - \eta_0^2$ and $B = 2 \eta_0^2 a$. Taking into account that

$$\sqrt{(A - jB)} = h + jg,$$

where

$$h = (1/\sqrt{2})[A + \sqrt{(A^2 + B^2)}]^{1/2};$$

$$g = -(1/\sqrt{2})[\sqrt{(A^2 + B^2)} - A]^{1/2} \qquad (4.34)$$

the expression for the reflection coefficient becomes

$$V = \frac{b \cos \theta_1 - j (h + jg)}{b \cos \theta_1 + j (h + jg)}$$

$$= \frac{b \cos \theta_1 + g - jh}{b \cos \theta_1 - g + jh} \qquad (4.35)$$

The modulus of this expression is given by

$$|V|^2 = \frac{(b \cos \theta_1 + g)^2 + h^2}{(b \cos \theta_1 - g)^2 + h^2} = a_r \qquad (4.36)$$

and is always less than unity since g is negative, so total internal reflection cannot take place, no matter what the angle of incidence may be. The phase-shift between the incident and reflected pressure waves is given by the argument of eqn. (4.35) as

$$\phi = \tan^{-1}\left(\frac{-h}{b \cos \theta_1 + g}\right) - \tan^{-1}\left(\frac{h}{b \cos \theta_1 - g}\right) \qquad (4.37)$$

It is interesting to look at the form of the transmitted wave under these conditions. The acoustic pressure wave in the second medium is given by eqn. (4.3) as

$$p_t = W\hat{p} \exp [j(\omega t - k_2 x \cos \theta_2 - k_2 y \sin \theta_2)]$$

Using Snell's Law and eqn. (4.26) the term $k_2 \cos \theta_2$ in this expression can be rewritten as $k_1\sqrt{(\eta^2 - \sin^2 \theta_1)}$, therefore

$$k_2 \cos \theta_2 = jk_1 \sqrt{(\sin^2 \theta_1 - \eta^2)}$$

From eqn (4.9)

$$W = 1 + V$$

therefore the expression for the transmitted wave becomes

$$p_t = \hat{p}(1 + V) \exp [k_1 x \sqrt{(\sin^2 \theta_1 - \eta^2)}]$$
$$\exp [j(\omega t - k_1 y \sin \theta_1)] \qquad (4.38)$$

This equation represents an inhomogeneous wave propagating in the y-direction, characterized by the final term of the expression which is attenuated exponentially in the direction normal to the boundary and the propagation direction with an attenuation coefficient,

$$a_x = k_1 \sqrt{(\sin^2 \theta_1 - \eta^2)}$$

The difference between reflection under ideal and real conditions is illustrated by Fig. 4.3 for the case $b = \rho_2/\rho_1 = 2 \cdot 0$, $\eta = 0 \cdot 9$, which corresponds to the reflection of an acoustic wave incident from sea-water upon a plane ocean floor of water-packed silt.

If the sea-floor absorbs no acoustic energy, total internal reflection occurs for all angles of incidence greater than 64° as

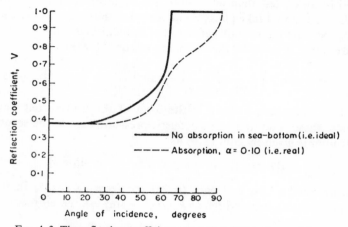

FIG. 4.3. The reflection coefficient at a water-silt (sea-bottom) interface under ideal and real conditions.

shown by the full line. When, however, the absorption of the sea-floor is taken into consideration by assuming a value of $a = a'/k = 0.10$, the reflection coefficient, as shown by the broken line, is always less than unity.

3. Transmission between Liquid and Solid Media

The reflection of acoustic waves from a boundary between a liquid and a solid is considerably more involved than that produced at the boundary between two liquids because the solid medium is able to sustain transverse waves in addition to longitudinal waves. Any wave incident from the liquid onto the boundary at an oblique angle can be resolved into components, normal to and tangential to the boundary which give rise to longitudinal and transverse waves respectively.

Analytically, the problem is similar to the one considered previously in that a system of equations can be used to represent the incident and reflected waves in the liquid medium and the longitudinal and transverse transmitted waves in the solid medium. These can then be solved by use of the boundary conditions which require that the normal components of particle displacement and stress be continuous across the boundary between the liquid and the solid and that the tangential component of stress be zero at the boundary since the liquid is unable to sustain transverse waves.

The derivation of the expressions for the reflection and transmission coefficients is outside the scope of a book concerned with underwater acoustics; the reader is referred therefore to the work of Brekhovskikh[1] which lays out the analysis in detail.

4. Refraction of Acoustic Waves: Velocity Profiles

The simplest form of refraction, i.e. the refraction of a plane acoustic wave at a plane interface between two discrete homogeneous media has already been considered, but within the volume of the sea a different type of refraction occurs because the sea is far from homogeneous, being subject to both regular and and random changes in its acoustic properties. Propagation anomalies resulting from the random changes will be discussed in Section 9, the effects of regular changes being discussed now.

These regular changes take the form of variations in the velocity of propagation of sound from place to place and particularly with depth. Equation (1.1) shows that the propagation velocity is influenced by three major factors: depth, temperature and salinity. The increase in velocity due to depth alone is quite regular and equals $0 \cdot 018$ m/sec per metre of depth, which is relatively insignificant. Salinity variations are also of little importance except perhaps near the surface of the sea, where the effects of evaporation are experienced, or near river estuaries; they result in an increase in velocity of 1·14 m/sec for an increase in salinity of one part per thousand. Velocity variations due to temperature, resulting in a velocity increase of approximately 0·966 m/sec°C rise in temperature are, however, quite considerable, and near the surface of the sea are very much dependent upon the latitude, the general weather conditions, the

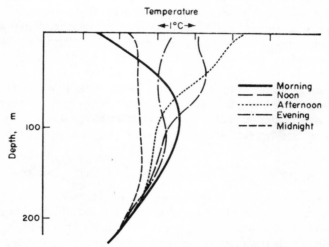

FIG. 4.4. Typical diurnal variation of temperature in the surface layer of the ocean.

season of the year and particularly the time of day. If the temperature of the ocean is plotted as a function of depth, a *thermal profile* is obtained. Diurnal variations in the shape of a typical thermal profile are shown in Fig. 4.4.

In the surface layer, which extends to a depth of approximately 200 m, sea-wave motion often mixes the water sufficiently for the temperature to be constant; the layer is then known as an *isothermal layer* and the variation of sound velocity with depth is due solely to the depth component.

Below about 200 m temperature variations become more regular, the temperature decreasing slowly with depth in what is known as a *thermocline* until at around 1000 m a temperature of 4°C is experienced. The temperature then remains constant with depth, a deep isothermal layer, until the sea-floor is reached. A typical

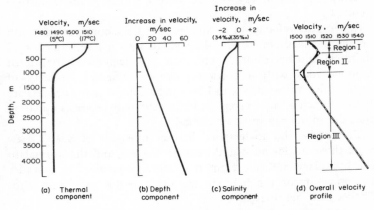

FIG. 4.5. A typical velocity profile showing its components. (Examples of ray tracing in Section 6 are based upon profile (d).)

velocity profile is shown in Fig. 4.5 together with its thermal, pressure and salinity components; this is not necessarily typical of the ocean in general, however, owing to the geographical and temporal variations experienced.

5. Ray Tracing

In the study of the effects of refraction due to non-uniform velocity of propagation it is very convenient to make use of the concept of a *ray*. This is a familiar concept, being used in elementary optics, and may be defined as a line drawn in the direction of propagation so that it is everywhere perpendicular to the wavefront. The computation of the actual line, and hence of the wavefront at different points in the medium, is a process known as *ray tracing*. Before considering the general problem of ray tracing in the ocean, however, it is necessary to investigate the path of a ray in a medium with a constant change of velocity with depth, i.e. a constant velocity gradient. It will be assumed that the sea possesses a vertical velocity gradient only, consequently, the sound ray will lie wholly within a single vertical plane.

When the velocity of propagation varies continuously with depth, the medium can be divided up into a large number of infinitely thin horizontal layers, each of which can be considered homogeneous with a constant, though different, velocity of propagation within it. Snell's Law can be applied to the boundaries between each of these layers and it immediately becomes apparent that the sound ray must be curved.

Previously, the angles used in the statement of Snell's Law were those angles measured between the ray and the normal to the boundary but at present it is more convenient to consider the angles measured between the ray and the horizontal, thus Snell's Law becomes

$$\frac{\cos \phi_1}{c_1} = \frac{\cos \phi_2}{c_2} \tag{4.39}$$

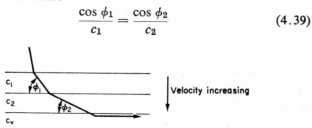

FIG. 4.6. Geometry of a sound ray propagating through a number of infinitely thin layers each with a constant propagation velocity.

where c_1 and c_2 are the velocities in two adjacent layers and ϕ_1 and ϕ_2 are the angles of inclination of the ray within them, as shown in Fig. 4.6, where it is assumed, as often happens, that velocity increases with depth.

Provided the medium is sufficiently extensive, and the velocity continues to increase with depth, one particular depth is reached at which the ray becomes horizontal; the angle of inclination is then zero and the corresponding velocity of propagation is known as the *vertex velocity* c_v. This leads to a more convenient form for Snell's Law:

$$\frac{\cos \phi_p}{c_p} = \frac{1}{c_v} \qquad (4.40)$$

where c_p and ϕ_p are associated with any layer within the medium.

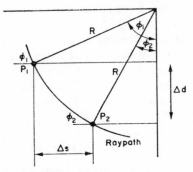

Fig. 4.7. Ray path geometry in a medium with a constant velocity gradient.

When the velocity gradient is constant the sound path is an arc of a circle as shown in Fig. 4.7. In this, the radius of curvature R of the ray is related to the difference in depth Δd of two points P_1 and P_2 on the ray, thus:

$$\Delta d = R (\cos \phi_2 - \cos \phi_1) \qquad (4.41)$$

The velocities of propagation at P_1 and P_2 are related by

$$c_2 = c_1 + g\Delta d$$

where g is the velocity gradient; therefore

$$\Delta d = \frac{c_2 - c_1}{g} = R (\cos \phi_2 - \cos \phi_1)$$

Using eqn. (4.40) c_1, c_2, ϕ_1 and ϕ_2 can be eliminated from this expression to give the radius of curvature of the ray

$$R = \frac{c_v}{g} = \frac{c_p}{g \cos \phi_p} \tag{4.42}$$

For underwater applications it is usual to consider a velocity gradient as positive when the velocity of propagation *increases* with depth; the radius of curvature of the ray is also positive and the ray is bent upwards. When the velocity gradient is negative, the radius of curvature is negative also and the ray bends downwards.

Using the radius of curvature given by eqn. (4.42) the path of an actual ray can be computed using a few fundamental trigonometrical relationships. The horizontal distance Δs travelled by the sound ray while it changes its depth by the amount Δd is given by the horizontal distance between points P_1 and P_2 of Fig. 4.7.

$$\Delta s = R (\sin \phi_1 - \sin \phi_2) \tag{4.43}$$

Eliminating R between eqns. (4.41) and (4.43) gives

$$\Delta s = \Delta d \, \frac{\sin \phi_1 - \sin \phi_2}{\cos \phi_2 - \cos \phi_1}$$

i.e.

$$\Delta s = \Delta d \, . \, \cot \frac{(\phi_1 + \phi_2)}{2} \tag{4.44}$$

6. Ray Tracing in the Deep Ocean

Before proceeding to particular instances of ray tracing it should be pointed out that ray theory is applicable only when the scale of the parameters involved, e.g. range, depth of water, target size etc., is large compared with the acoustic wavelength. If this condition is not fulfilled then *diffraction* theory† must be employed. For the cases examined in this book, however, it is obvious that ray theory will suffice.

† For a discussion cf diffraction theory the reader is referred to the works of Rayleigh[4] and Lamb.[5]

The results of eqns. (4.39) to (4.44) which apply to media with constant velocity gradients only can be used to compute the path of a sound ray in the ocean where the velocity gradient changes continuously with depth, if the velocity profile is divided up into a number of discrete regions each of constant but different velocity gradient. Examination of many velocity profiles reveals that most practical velocity profiles can be approximated by as few as five such regions and for many purposes even fewer may suffice. This process when applied to the thermal profile of Fig. 4.5, as indicated by the broken line, yields three such discrete regions.

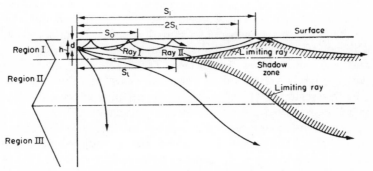

FIG. 4.8. The ray diagram corresponding to the velocity profile of Fig. 4.5.

If a sound source is located in any of these regions, say region I, it is possible to compute the path of every radiated sound ray. In practice, only the paths which delineate the illuminated volume of the sea are of any real interest, consequently the volume of computation required is small. One important path is that of a ray leaving the source horizontally. Under the conditions of Fig. 4.5 this will be bent upwards and will strike the surface at a horizontal distance

$$s_0 = \sqrt{[(2R - d)\,d]}$$

where d is the depth of the source, as shown in Fig. 4.8. However, $R \gg d$ so this expression reduces to

$$s_0 = \sqrt{(2\ Rd)} \qquad (4.45)$$

This ray will be reflected from the surface in a downward direction; the positive gradient in region I, however, will cause it to be refracted upwards again and it will strike the surface once more. This process will be repeated and the ray will strike the surface at successive ranges given by the expression

$$s_n = (2n + 1)\ s_0$$

where $n = 0, 1, 2, 3$, etc.

The ray (ray II of Fig. 4.8) leaving the source in a downward direction at such an initial angle that it becomes horizontal at the boundary between regions I and II is of special importance. This ray, after grazing the boundary, undergoes upward refraction and eventually strikes the surface. The ray leaving the source with a slightly greater initial angle will cross the boundary and pass into region II where because of the negative velocity gradient it will be refracted downwards away from region I. The ray that grazes the boundary, termed a *limiting ray*, therefore defines the limits of a *shadow zone* into which energy radiated from the source cannot enter, except possibly by reflection from the surface of rays possessing other initial directions.

The horizontal distance between the source and the point at which the limiting ray grazes the boundary is given by

$$s_l = \sqrt{[2R_1\ (h - d)]} \qquad (4.46)$$

where h is the depth of the boundary between the two regions and R_1 is the radius of curvature of the sound ray in region I. The limiting ray after grazing the boundary again reaches the level of the sound source at a horizontal distance of $2\ s_l$.

When the limiting ray becomes horizontal at the grazing point, ϕ_2 in eqn. (4.41) is zero and this equation becomes

$$\Delta d = R_1 (1 - \cos \phi_1) \qquad (4.47)$$

where Δd is the height of any point in the limiting ray above the boundary between the two regions. At the sound source $\Delta d = h - d$; therefore the initial angle ϕ_1 of the limiting ray is given by the expression

$$\cos \phi_1 = 1 - \frac{h - d}{R_1} \qquad (4.48)$$

The expressions obtained so far enable a few specific points on the limiting ray path to be determined quickly, but the computation of the complete path of this ray and all other rays through the various regions is a far more tedious operation.

The general procedure consists of determining the horizontal range from the source of each sound ray as it passes through certain increments of depth spaced throughout the various regions. The radius of curvature of the sound ray varies from region to region and its value R_n in each region must first be determined from the relevant velocity profile. Consider a ray leaving the source with an angle of inclination ϕ_n; at a depth d_n its angle of inclination ϕ_m will be given by eqn. (4.41) re-written in the form

$$\cos \phi_m = \cos \phi_n + \frac{d_n - d}{R_n} \qquad (4.49)$$

(In this expression the sign of R_n, which may be negative, must be taken into account.)

The horizontal distance between the source and the point on the ray corresponding to this depth d_n is given by eqn. (4.43) as

$$s_n = R_n (\sin \phi_n - \sin \phi_m) \qquad (4.50)$$

If this process is repeated for a number of different depths the path of the ray can be traced out. When the ray reaches the boundary between two regions the value of R_n will alter and in a large number of cases its sign will be reversed also, and these two factors must be taken into account in eqns. (4.49) and (4.50). It should be noted that each ray has a different radius of curvature from its neighbours in each region depending upon its initial angle as it leaves the source.

6.1. *Sound Channels*

When the sound source is located in region II of Fig. 4.5 the ray path configuration is entirely different as shown in Fig. 4.9. Once again, a limiting ray marking the limits of a shadow zone is experienced. Rays leaving the source at angles (measured from the vertical) less than that of the limiting ray pass into region I and never return to region II. Rays leaving with greater angles than that of the limiting ray are refracted downwards into region

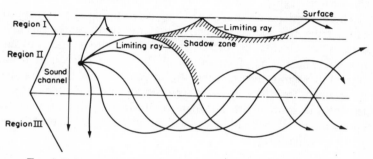

Fig. 4.9. A ray diagram showing the formation of a sound channel.

III. Here the positive velocity gradient causes them to undergo upward refraction and they eventually pass back into region II. Once again they are refracted downwards and thereafter they continually pass between regions II and III until they are attenuated. The minimum in the velocity profile therefore attracts or channels the sound waves to its own level and for this reason is known as a *sound channel*.

Sound waves constrained to travel in such a channel do not diverge with spherical spreading as they would in isovelocity (constant velocity) water; rather, the divergence is cylindrical. Low-frequency signals which suffer little absorption can therefore be expected to propagate to great distances under such conditions. As mentioned in Section 4 the typical deep-ocean velocity profile does possess a minimum at a depth of around 1000 m, and therefore *deep sound channels* are found in many parts of the ocean, some extending over distances in excess of 3000 km, which could prove useful in facilitating rescue work at sea. A technique using the deep sound channel, known as SOFAR (*SO*und *F*ixing *A*nd *R*anging) has been developed where anyone in distress detonates a small charge of explosive in the channel. The position of the explosion can then be determined by the differences in the transit times to three or more listening stations.

6.2. *Signal Attenuation Due to Refraction*

Even when sound rays are not constrained to a sound channel the variation of sound intensity as a function of range is found to depart noticeably from the inverse-square or spherical spreading

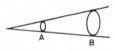

(a) Isovelocity conditions   (b) Conditions of constant velocity gradient

FIG. 4.10. The change of area of a tube of acoustic energy under different conditions.

law. This is illustrated by Fig. 4.10 which compares the cross-sectional areas of tubes of acoustic energy formed by a number of sound rays starting out with only slightly different initial directions under isovelocity and velocity-gradient conditions. In the two cases the ratios of the cross-sectional areas and hence the acoustic intensities are widely different.

It has been shown by Horton[3] that the acoustic intensities in a sound tube of this sort in a medium with a constant velocity gradient are related by the expression

$$\frac{I_A}{I_B} = \left(\frac{s_B}{s_A}\right)^2 \qquad (4.51)$$

at two points A and B where I represents acoustic intensity and s_A and s_B are the horizontal components of distance from the source and not the total path lengths.

The determination of acoustic intensity at a point in the ocean using ray tracing techniques is outside the scope of this book but a complete account of the procedure is given by Horton.

7. Ray Tracing in Shallow Water

In shallow water each sound ray will undergo many reflections from the surface and sea-bed and account must be taken of the phase changes occurring at each reflection. There will be a discontinuous increase in velocity across the sea-floor as the sound ray passes from the lower velocity of the sea to the higher velocity of the sediment. The reflection coefficient depends upon the angle of incidence of the ray, being less than unity for all angles between the normal and the critical angle, and unity or very nearly so for angles of incidence between the critical angle and the horizontal. Thus only rays starting out with angles to the vertical greater than the critical angle would be expected to propagate to great distances. The phase change between the incident and reflected waves, which is independent of frequency, is zero for all angles of incidence less than the critical angle but increases regularly from $0°$ to $180°$ between the critical angle and the horizontal.

There is a large change in acoustic properties at the surface of the sea which results in a soft boundary at this point; so the resultant pressure on the surface must be zero, since, otherwise, the boundary would possess infinite acceleration.

The reflection coefficient is very nearly unity and the phase shift is $180°$.

The direction and intensity of sound rays in shallow water can be determined without taking account of the phase changes on reflection but, as will be seen later, because of these, only waves of certain frequencies can actually be propagated along a specific ray path.

7.1. *Ray Tracing Under Isovelocity Conditions*

In an area where no change in sound velocity, either laterally or vertically, takes place, all rays will be straight lines and simple geometry gives the paths of all the rays originating from a given sound source. Waves generated by a point source will be spherical waves and their intensity will fall off as the square of the total distance travelled.

Owing to the many different propagation paths existing between any two points, the received signal is extremely complex as shown in Fig. 4.11 (a) for a source and receiver at the surface in a region where the sea-floor and surface are plane and parallel. Ray 1 starting out with an angle of incidence (relative to the vertical) less than the critical angle is heavily attenuated by successive reflections where much of the energy is transmitted into the sea-floor, and is of very low intensity at the receiver.

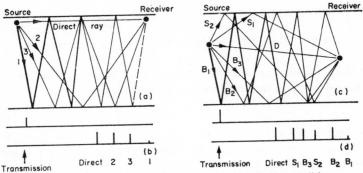

FIG. 4.11. Shallow water ray diagrams under isovelocity conditions: (a) ray diagram when source and receiver at surface, (b) corresponding signals at the receiver for a single impulse transmitted, (c) ray diagram when source and receiver in midwater, (d) corresponding signals at the receiver for a single impulse transmitted.

Rays 2 and 3 having initial angles of incidence greater than the critical angle are attenuated only by absorption within the volume of the sea and are far more intense. The time of arrival of each impulse at the receiver is dependent upon the total distance travelled.

If the source and receiver are located in midwater, the received signals become even more complex because rays initially incident upon the surface must be included also. A typical sequence of ray paths and their corresponding received signals are shown in Fig. 4.11 (c) and (d).

The ray paths shown in Fig. 4.11 are only a few of the infinity of paths existing between source and receiver. Under practical conditions, however, phase changes occurring at the upper and lower boundaries render most of these paths untenable.

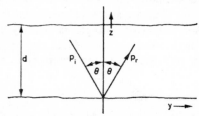

FIG. 4.12. Plane waves incident obliquely upon a plane sea-floor.

Referring to Fig. 4.12 the acoustic pressure wave incident at an angle θ upon the sea-floor can be expressed by

$$p_i = \hat{p} \exp\left[j(\omega t + kz \cos \theta - ky \sin \theta)\right] \qquad (4.52)$$

Assuming the sea-floor to be perfectly rigid, the reflection co-efficient V will be unity and the reflected wave can be expressed by

$$p_r = \hat{p} \exp\left[j(\omega t - kz \cos \theta - ky \sin \theta)\right] \qquad (4.53)$$

The total pressure within the sea is, therefore:

$$p = p_i + p_r$$
$$- 2\hat{p} \exp\left[j(\omega t - ky \sin \theta\right] . \cos (kz \cos \theta) \qquad (4.54)$$

If this wave is to propagate the conditions at the sea-surface must be satisfied also. Since this is a soft boundary the pressure must vanish at $z = d$. This can be achieved only if

$$\cos (kd \cos \theta) = 0$$

i.e. if

$$kd \cos \theta = (2n - 1) \, \pi/2 \qquad (4.55)$$

Substitution for k gives

$$\cos \theta = (n - 1/2) \lambda/2d \qquad (4.56)$$

This equation defines the initial angles of incidence measured from the vertical of all the rays propagating along tenable paths.

In reality, although the surface of the sea approaches, very nearly, the ideal soft boundary, the boundary between the sea and the sediment of the sea-floor is often far from rigid and account has to be taken of the phase shift occurring there. Under conditions of total internal reflection, this phase shift is given by eqn. (4.31) as

$$\phi = - 2 \tan^{-1} \frac{\sqrt{(\sin^2 \theta - \eta^2)}}{b \cos \theta}$$

therefore eqn. (4.55) can be rewritten as

$$kd \cos \theta - \phi = (2n - 1) \pi/2$$

whence

$$\cos \theta = (n + \phi/\pi - 1/2) \lambda/2d \qquad (4.57)$$

Unfortunately, the phase shift ϕ in this expression is also a function of the angle of incidence θ, consequently a general solution for θ is not readily obtainable and eqn. (4.57) must be solved graphically.

The problem can be further generalized to take account of the absorptive nature of the sea-floor by using the value of ϕ given by eqn. (4.37).

7.2. Ray Tracing in Regions with Velocity Gradients

When variations of sound velocity with depth are taken into account a still more general solution is possible.

Under conditions of negative velocity gradients the sound rays are refracted downwards and the range of reception of rays undergoing a large number of reflections is reduced by losses incurred at each reflection. Some rays which under isovelocity conditions would be first reflected from the surface now undergo their first reflection from the sea-floor. Confusion in the resulting ray diagram may be avoided by separating the direct rays and rays reflected from the surface, from the rays reflected from the sea-floor as shown in Fig. 4.13 (a) and (b). Ray 3 represents a ray which has become tangential at the surface and so forms a limiting ray delineating a shadow zone into which sound due to

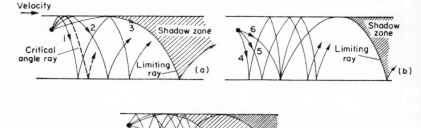

FIG. 4.13. Ray tracing in shallow water: (a) surface-reflected rays, (b) bottom-reflected rays, (c), (a) and (b) combined.

direct and surface-reflected rays alone cannot penetrate. Ray 1 meets the bottom at an angle of incidence less than the critical angle and is thus severely attenuated with range. Ray 6 represents a ray which after reflection from the sea-floor becomes tangential at the surface and is therefore a limiting ray defining another shadow zone in which no energy due to bottom-reflected rays can appear.

When these two figures are combined as in Fig. 4.13 (c) the complete picture is obtained and it is found that sound energy does appear in both shadow zones due to rays which arrive by reflection from the boundary other than the one producing the shadow zone. Once again, the requirement that pressure must vanish at the sea-surface restricts the number of tenable ray paths.

Under conditions of positive velocity gradient all sonud waves are refracted upwards and only surface-reflected rays are of any consequence. A typical ray path diagram under such conditions is shown in Fig. 4.14. The broken line in this diagram is the envelope to the family of ray paths and is known as a *caustic*. This caustic marks the limits of another shadow zone. In certain areas, as, for example, the shaded area in the diagram, sound rays are concentrated together and sound intensities considerably

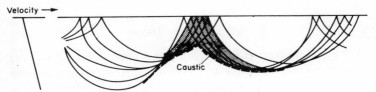

FIG. 4.14. Focusing by reflection under conditions of positive velocity gradient.

higher than would be predicted using the inverse-square law are observed. This effect, known as *focusing by reflection*, is of little practical importance because the positions of the regions of high intensity vary so rapidly as the velocity profile alters.

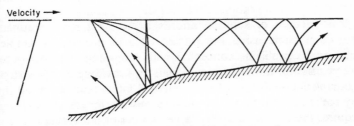

FIG. 4.15. Ray tracing in regions of sloping sea-floor.

7.3. *Ray Tracing over Sloping Sea-floors*

Long-range, shallow-water propagation is usually associated with the seas over a continental shelf. Such shelves generally slope gently away from the land mass and the velocity profile changes as the depth of the water changes; the ray diagram is thus made more complex to determine.

A typical ray diagram under these conditions is shown in Fig. 4.15 where downward refraction is assumed to occur.

Local changes of slope greatly affect the path of these rays and the problem of ray tracing under these extreme conditions is a very complex one which is not usually attempted.

8. Normal Mode Theory

Although ray tracing is adequate for a large number of sound propagation investigations it is worthwhile to mention that solutions to certain types of sound transmission problems can be obtained more readily in terms of normal modes. A normal mode defines for a given frequency a mode of vibration of a system which in this case consists of the sea and its boundaries; and a solution derived in terms of normal modes entails a summation of the contributions from these various preferred modes. A normal-mode solution is therefore a solution of the general wave equation, expressed in cylindrical co-ordinates, obtained using the boundary conditions applying to a particlar sound source and configuration of sea-surface and sea-floor.

This method of solution affords a more rigorous treatment than is possible using ray tracing. It also provides a simpler solution to problems concerned with long-range shallow-water propagation of low-frequency signals by providing in a few steps a complete description of the combined interference from all the ray paths incident upon the point of observation. A ray solution requires the tracing of a very large number of individual rays for an adequate description of the resulting interference; this is very tedious.

The subject of normal mode solutions is far too complex to be treated in this book, and for a complete description of the method the reader is advised to consult the works of Officer and Pekeris.[6, 7]

9. Random Effects in Acoustic Propagation in the Sea

So far in this chapter we have been able to assume that the parameters of the system have been known, and however complicated the spatial distribution of temperature, velocity, etc., it was at least possible to specify it. On this basis the acoustic propagation has been calculable, even if great labour would have been required to obtain complete results in many cases. But in practice there is another group of effects which is quite different in that the variations within the medium cannot be specified in regard to space or time; they are quite random. Some of these effects were mentioned in Chapter 1, particularly reverberation and fluctuation. The former affects echo-location systems by providing a background of random noise produced by the back-scattering of the signal from all the random inhomogeneities in the water, and where relevant also those on the sea-bottom or surface. The latter causes signal waveforms to be spoilt by the randomly varying propagation losses due to such effects as turbulence in non-isothermal water, aeration, and random phase of combination of rays which have travelled over different paths with reflections from unsteady surfaces such as large sea-waves.

Random effects of this kind can be specified only in terms of statistical parameters, as discussed in relation to noise in Chapter 2. Although a good deal of research has been done on these topics, it still remains more or less impracticable to make reliable predictions, and designers are forced to use a good deal of empirical data and to rely heavily on experience. A fuller discussion of the matter would be out of place in this book, but reference may be made to numerous research papers[8, 9, 10, 11] and certain handbooks[12, 13] if further information is required.

References

1. BREKHOVSKIKH, L., *Waves in Layered Media*, Academic Press, London (1960).
2. SCHELKUNOFF, S. A., *Electromagnetic Fields*, Blaisdell Publishing Co., New York (1963).

3. HORTON, J. W., *Fundamentals of Sonar*, United States Naval Institute, Annapolis (1957).

4. RAYLEIGH, LORD *Theory of Sound*, 2, Dover Publications, New York (1945).

5. LAMB, H., *The Dynamical Theory of Sound*, Dover Publications, New York (1960).

6. OFFICER, C. B., *Introduction to the Theory of Sound Transmission with Applications to the Ocean*, McGraw-Hill, New York (1958).

7. PEKERIS, C. L., *Theory of Propogation of Explosive Sound in Shallow Water*, Geol. Soc. America Mem. 27. 1948.

8. CLAY, C. S., Fluctuations of Sound Reflected from the Sea Surface, *J. Acoustical Soc. America* 32, 1547 (1960).

9. MINTZER, D., Wave Propagation in a Randomly Inhomogeneous Medium I and II, *J. Acoustical Soc. America* 25, 922 and 1107 (1953).

10. WHITMARSH, D. C., SKUDRZYK, E., URICK, R. J., Forward Scattering of Sound in the Sea and its Correlation with the Thermal Microstructure, *J. Acoustical Soc. America* 29, 1123 (1957).

11. KNAUSS, J. A., An Estimate of the Effect of Turbulence in the Ocean on the Propagation of Sound, *J. Acoustical Soc. America* 28, 443 (1956).

12. CHERNOV, L. A., *Wave Propagation in a Random Medium*, McGraw-Hill, New York (1960).

13. ALBERS, V. M., *Underwater Acoustics*, Plenum Press, New York (1963).

CHAPTER 5

Transducers for Underwater Use

1. General

The name *transducer* is given to devices which convert energy from one form to another; and to convert acoustical signals into electrical signals suitable for processing, *electroacoustic* transducers are used. It is usual to use electroacoustic transducers for the generation of underwater sound also, those used solely for this purpose being called *transmitters*. Transducers used solely for reception are called *hydrophones*. Transmitters and hydrophones often differ in their methods of operation but it is very common for the same transducer or same type of transducer to perform the tasks of both transmission and reception, e.g. transducers used for sonar and communication systems. These latter must possess the characteristics of *linearity* and *reversibility*. A transducer is linear if it produces an exact equivalent in electrical terms of the incident acoustic waveform, and it is reversible if it has the ability to convert energy between the electrical and acoustical forms in either direction. The term reversible is frequently used, however, in a more specialized form which denotes that energy is converted with equal efficiency in either direction. In this book we shall be largely concerned with transducers possessing these two properties.

The majority of transducers fall into two categories, those which employ electric fields in their transduction process and those which employ magnetic fields. Some are inherently linear whilst others have to be *polarized* to produce linear action. This arises from the fact that the force producing acceleration of the active mass of the transducer, which in turn causes acoustic radiation, can be directly proportional to or proportional to the square of the applied electrical signal, depending upon the physical mechanism employed for transduction. If the electrical signal is sinusoidal of the form

$$E(t) = E_e \sin \omega t$$

a square-law transducer produces a force proportional to

$$(E_e \sin \omega t)^2 = \tfrac{1}{2} E_e^2 (1 - \cos 2 \omega t)$$

i.e. a steady component plus a sinusoidal component varying at twice the applied frequency. If now a steady polarizing quantity E_0 is applied together with the sinusoidal signal the force produced is modified, becoming proportional to

$$(E_0 + E_e \sin \omega t)^2 = E_0^2 + 2 E_e E_0 \sin \omega t + E_e^2 \sin^2 \omega t$$

and if E_0 is made much greater than E_e the final term of this expression can be ignored and an alternating force is produced which is a linear function of the applied sinusoid.

TABLE 5.1. EXAMPLES OF ELECTRO-ACOUSTIC TRANSDUCERS

Electric field types			
Type	Category	Frequency range	Examples
Piezoelectric	Linear	5 kc/s–50 Mc/s	Crystal microphones Transmitters and hydrophones Shear wave generators
Dielectric	Polarized	Up to 100 kc/s	Condenser microphones Electrostatic loudspeakers
Electrostrictive	Polarized	5 kc/s–10 Mc/s	Transmitters and hydrophones Bimorphs for microphones and loudspeakers Shear wave generators
Spark source	Irreversible	High frequency Broad-band	High intensity transmitter

Table 5.1—*continued*

Magnetic field types

Type	Category	Frequency range	Examples
Electrodynamic[1]	Linear	Up to 50 kc/s	Moving coil micro-phones and loud-speakers
Electromagnetic[2]	Polarized	Up to 50 kc/s	Moving iron micro-phones and earpieces
Magnetostrictive	Polarized	Up to 200 kc/s	Transmitters and hydrophones
Dynamometric[3]	Irreversible	Low frequency Broad band	High intensity transmitter

Active receivers

Type	Category	Frequency range	Examples
Carbon micro-phone	Irreversible	Up to 50 kc/s	Telephone transmitter
Thermal	Irreversible	Up to 600 kc/s	Hot-wire microphone

[1] Electrodynamic transducers utilize the force on a current carrying conductor in a magnetic field to vibrate a radiating diaphragm.

[2] Electromagnetic transducers employ electromagnets, the coils of which carry a varying electric current to attract and repel their diaphragms.

[3] In the dynamometric transducer the magnetic fields of two coils carrying currents in the same sense interact and force the coils apart.

A square-law transducer if unpolarized cannot be used as a receiver. A field, either electric or magnetic, must first be present, which vibrations within the transducer can alter, before an electrical output can be produced. Such a transducer must therefore be polarized to make it both linear and reversible.

A few of the more commonly employed types of transducer are indicated in Table 5.1 together with a few of their properties. Of these only the piezoelectric, the magnetostrictive and the electrostrictive types (in that chronological order) have ever found widespread applications underwater although a few others are used for specialized purposes, e.g. high intensity sound sources, geophones for seismic studies, pressure transducers, etc.

2. Piezoelectric Transducers

The direct and converse piezoelectric (pressure-electric) effects are exhibited by certain crystals in which separation of the "centres of gravity" of their positive and negative charges occurs as a result of mechanical stress. Such a crystal is said to be asymmetrical along the axis normal to the stress. This charge separation produces electrical dipoles within the crystal which in turn induce surface charges. Crystals vary considerably in the number of axes of asymmetry they possess and transducer elements have to be cut with their faces parallel to such an axis.

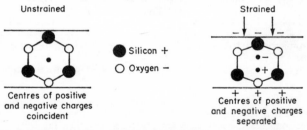

FIG. 5.1. Piezoelectric action in quartz showing separation of charge centres under stress.

This direct piezoelectric effect is illustrated in Fig. 5.1. When the crystal is in an unstrained state the centres of the positive and negative charges are coincident. When the slice is compressed, however, these centres move relative to one another and surface charges are produced. Provided the crystal is not strained beyond its elastic limit the magnitude of the charge density is proportional to the applied stress. The converse effect also exists; if an electric field is applied along an axis of asymmetry, the slice changes its physical dimensions, the amount of strain being proportional to the electric field intensity.

Piezoelectricity is found in many crystals, the effect being most strongly pronounced in the following materials, quartz; Rochelle salt, lithium sulphate and ammonium dihydrogen phosphate (ADP). The behaviour of piezoelectric materials is characterized by the relationships described below.

2.1. *Piezoelectric Relationships*

Consider first the direct piezoelectric effect. The surface charge induced on a slice of piezoelectric material is called the *polarization charge P* and is measured in coulombs/m². It is related to an applied stress T, measured in newtons/m² by

$$P = Td \qquad (5.1)$$

where d is the *piezoelectric strain coefficient*, defined as the charge density per unit applied stress, measured in coulombs/newton, when no external electric field is applied to the slice, i.e. under short-circuit conditions.

If now an electric field E measured in V/m, is applied to the slice, the electric flux density D, in coulombs/m², within it becomes

$$D = E\epsilon + Td \qquad (5.2)$$

where ϵ is its permittivity, measured in farads/m.

Now consider the converse effect. If an unstressed slice is subjected to an electric field it undergoes a mechanical strain S which is related to the electric field intensity by

$$S = Ed' \qquad (5.3)$$

Consideration of the principle of the conservation of energy shows that $d = d'$ and yields an alternative definition for d, i.e. the mechanical strain produced per unit applied field, measured in m/V under conditions of no load.

Applying a tensile stress T to the slice which possesses an elastic constant s, measured in m²/newton, results in a total strain

$$S = Ts + Ed \qquad (5.4)$$

In a non-piezoelectric material eqns. (5.2) and (5.4) reduce to the well known relations

$$D = \epsilon E \qquad (5.5)$$

$$S = Ts \qquad (5.6)$$

When a compressive stress $-T$ is applied, eqn. (5.4) becomes

$$S = -Ts + Ed \qquad (5.4a)$$

and the slice can be effectively clamped if S is made zero by balancing the strain produced by the electric field by the compressive strain. Under these conditions eqn. (5.4a) gives

$$T = eE \qquad (5.7)$$

where $e = d/s$ is the *piezoelectric stress coefficient* defined as the stress produced per unit applied field 'measured in newtons/Vm.

Equations (5.1) and (5.6) give

$$P = eS \qquad (5.8)$$

and an alternative definition for e, i.e. the charge density obtained per unit strain, expressed in coulombs/m². These simple relationships, eqns. (5.1) to (5.8), apply to one direction only, i.e. the thickness direction of the slice. A stress may be applied, however, in any direction relative to the slice when it must be resolved into six components, three tensile, T_1, T_2, T_3, along the x-, y- and z-axes and three shear, T_4, T_5, T_6, along the yz-, xz- and xy-planes. There will be six corresponding strains. Any resulting polarization charge and any external applied field must also be resolved into their x-, y- and z-components: P_x, P_y, P_z and E_x, E_y, E_z.

Thus eighteen piezoelectric stress and eighteen piezoelectric strain constants will be required to specify completely the piezoelectric relations and the simple relations of eqns. (5.3) and (5.6) must be replaced by matrices of stress–field and stress–strain relationships in the manner explained by Mason.[1]

Fortunately, no piezoelectric output occurs along axes of symmetry; this and considerations of symmetry reduce the number of the relevant constants. Taking quartz, for example, all d-coefficients are zero except

$$d_{11}, d_{12}, d_{14}, d_{25} \text{ and } d_{26}$$

and because of symmetry

$$d_{12} = - d_{11}, \; d_{25} = - d_{14}, \; d_{26} = - 2d_{11}$$

Thus the piezoelectric relations for quartz simplify to

$$
\begin{aligned}
S_1 &= d_{11} E_x & S_4 &= d_{14} E_x \\
S_2 &= - d_{11} E_x & S_5 &= - d_{14} E_y \\
S_3 &= 0 & S_6 &= - 2d_{11} E_y
\end{aligned}
$$

where $d_{11} = - 2 \cdot 3 \times 10^{-12}$ and $d_{14} = 0 \cdot 57 \times 10^{-12}$ coulombs/newton.

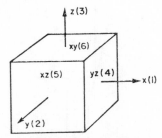

FIG. 5.2. Nomenclature of crystal axes and planes.

At this point it is necessary to explain the system of subscripts used for the various constants. Numbers 1, 2, 3, 4, 5, 6 refer to the x-, y-, z-axes and the yz-, xz- and xy-planes of Fig. 5.2 respectively. The first subscript of a piezoelectric constant denotes the direction of the applied field, the second the direction of the resulting strain. Thus d_{11} and d_{14} imply that in both cases the field is applied along the x-axis whilst the strains are compressional and shear strains along the x-axis and yz-plane respectively. The first subscript of an elastic constant denotes the direction of the applied stress, the second the direction of the resulting strain.

Underwater transducer design using piezoelectrics is greatly simplified as one is usually only concerned with longitudinal waves generated in the x-direction by a field also in the x-direction, the relevant constants having the subscript 11.

2.2. Piezoelectric Vibrators

Piezoelectric effects are exhibited along directions depending upon the class to which a particular crystal belongs and transducer elements have to be cut to take advantage of these directions. Discussion of the various ways of cutting a crystal will be illustrated by considering quartz—a material most commonly employed owing to its great physical and chemical stability.

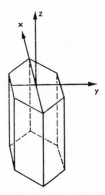

FIG. 5.3. The x, y and z axes of quartz.

A typical quartz crystal is shown in Fig. 5.3. In this, the Z-axis, the optical axis, is one of symmetry. The three X- and the three Y-axes join opposite edges and faces respectively. Slices cut with their faces normal to the X- or Y-axes exhibit piezoelectric properties and are known as X- or Y-cuts respectively.

When such a slice is coated with metal on opposite faces an electrical capacitor is formed. If an electric potential is then applied across this capacitor the resulting electric field, depending upon its polarity, causes an X-cut crystal to expand or contract in the X-direction. A strain of opposite sense to that occurring in the X-direction occurs simultaneously along the Y-direction but no strain occurs along the Z-direction. If the potential alternates the crystal must perform small-amplitude forced vibrations. If its frequency is varied the vibrations will be a maximum at frequencies corresponding to the natural frequencies of vibration of particular axes of the slice giving rise to a phenomenon known as *resonance*. These resonances can occur in length and either breadth or thickness modes, according to which is the Y-axis. The resonant frequencies are inversely proportional to the dimensions of the crystal along its vibration axes.

The crystal can also be made to resonate at harmonic frequencies corresponding to odd submultiples of the dimensions of the slab along its vibration axes. Vibrations of this type are restricted to odd harmonics only and the electroacoustic energy conversion efficiency decreases with increasing order of hamornic.

Similar arguments can be applied to Y-cut crystals. If the slab is rotated in the Z-direction before cutting, its properties are modified and its frequency stability with changes of temperature can be enhanced. This technique is also used to suppress unwanted modes of vibration.

3. Electrostrictive Transducers

All dielectric materials exhibit the phenomenon of electrostriction but in the class of dielectrics known as *ferroelectrics*, the effect is very pronounced. Within these materials electric dipoles are formed spontaneously which have a preferred orientation within certain localized regions or *domains*. Ordinarily, these domains are randomly disposed and the overall electric moment of the material is zero. Application of an electric field, however, causes the domains to become aligned with the field and the physical dimensions of the material alter. The mechanical strain is independent of the sense of the applied field, positive and negative fields producing the same strain; thus electrostrictive transducers are inherently non-linear and must be polarized to give true transducer action. This can be done permanently by heating the material to a temperature at which the electrostrictive properties disappear (the Curie temperature) and then allowing it to cool while subjecting it to a high-intensity electric field. Polarized electrostrictive materials have similar properties to piezoelectrics and are often referred to as piezoelectric. Converse electrostriction is displayed by polarized ferroelectrics.

3.1. *Electrostrictive Relationships*

Equations 5.1 to 5.8 and the system of coefficients and subscripts developed for piezoelectric materials can be used to describe the action of electrostrictive materials also. There is one major point of difference, however; the value of the permittivity ϵ is dependent, though not linearly, upon the intensity of the polarizing electric field E_0. Hence D, d and e must be specified for a particular value of E_0. When the material is used in single crystalline form, axes of symmetry need be avoided no longer and it is usual to apply the electric field and polarize the element along its Z-axis, the relevant subscript being 33.

3.2. *Electrostrictive Vibrators*

Ferroelectric materials are usually in the form of polycrystalline solids and are made into transducer elements by bonding them with suitable additives (which can also be used to modify the electrostrictive properties) to form a ceramic which is then moulded into the desired shape. The resulting element is anisotropic and does not have to be cut along specific axes thus permitting a multitude of shapes. Some of these are shown in Fig. 5.4, and include plates, discs, tubes, hollow spheres and concave bowls. Some of these shapes possess axes of circular symmetry and permit modes of vibration other than those possible with piezoelectrics. Thus a disc can execute either thickness or radial vibrations and a tube can vibrate in either its thickness, length or circumferential modes. In this latter mode the mean radius of the tube increases and decreases. The moulded ceramic is ground to the exact size corresponding to the required operating frequency.

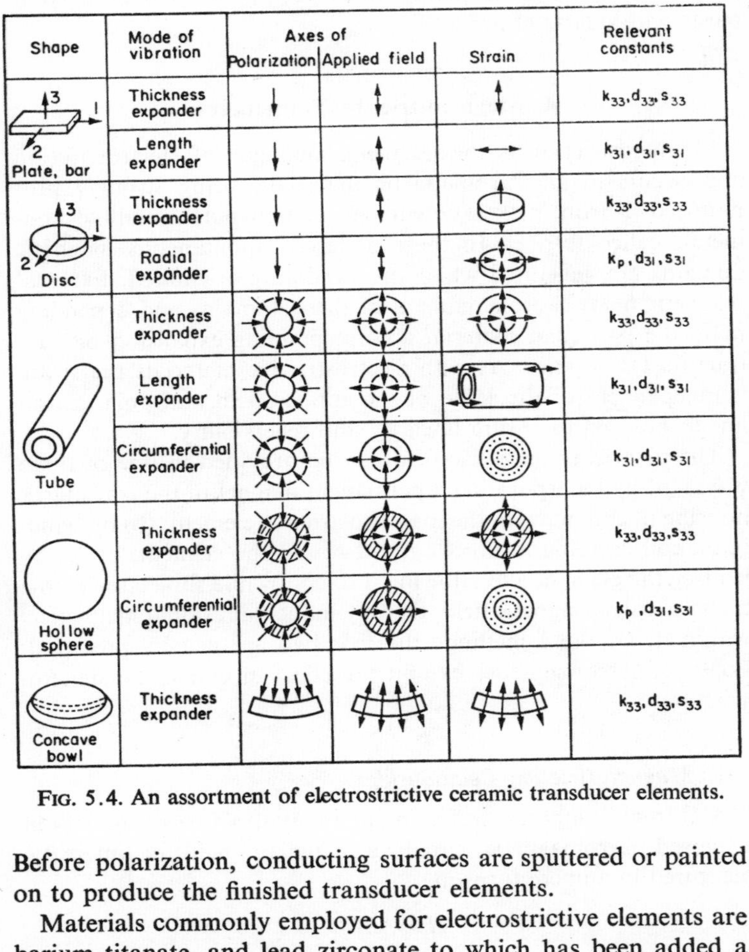

Shape	Mode of vibration	Axes of Polarization	Applied field	Strain	Relevant constants
Plate, bar	Thickness expander				k_{33}, d_{33}, s_{33}
Plate, bar	Length expander				k_{31}, d_{31}, s_{31}
Disc	Thickness expander				k_{33}, d_{33}, s_{33}
Disc	Radial expander				k_p, d_{31}, s_{31}
Tube	Thickness expander				k_{33}, d_{33}, s_{33}
Tube	Length expander				k_{31}, d_{31}, s_{31}
Tube	Circumferential expander				k_{31}, d_{31}, s_{31}
Hollow sphere	Thickness expander				k_{33}, d_{33}, s_{33}
Hollow sphere	Circumferential expander				k_p, d_{31}, s_{31}
Concave bowl	Thickness expander				k_{33}, d_{33}, s_{33}

FIG. 5.4. An assortment of electrostrictive ceramic transducer elements.

Before polarization, conducting surfaces are sputtered or painted on to produce the finished transducer elements.

Materials commonly employed for electrostrictive elements are barium titanate, and lead zirconate to which has been added a small amount of lead titanate. These ceramics are more strongly piezoelectric than quartz but possess inferior thermal stability. They also must be operated at lower temperatures, barium titanate and lead zirconate–titanate losing their transduction properties at Curie temperatures of 120°C and 320°C as compared with quartz at 573°C.

4. Magnetostrictive Transducers

Magnetostriction is the magnetic analogue of electrostriction and occurs in all ferromagnetic materials, being strongly pronounced in iron, nickel, cobalt and certain polycrystalline nonmetals called ferrites. In these materials spontaneous magnetic domains are generated which are randomly orientated. External magnetic fields cause alignment of these domains and depending upon the particular material a corresponding expansion or contraction takes place. As with electrostrictive materials the strain is independent of the sense of the applied field and polarization has to be used to ensure linearity and reversibility.

The polarizing magnetic field can be provided in one of three ways: (a) by incorporating a permanent magnet in the transducer housing in contact with the magnetostrictive element, (b) by winding a coil carrying a direct current round the element, or (c) by heating the element and allowing it to cool while subjecting it to a high intensity magnetic field, thereby inducing residual magnetism within it. Of these methods the first two are usually preferred. Polarized ferromagnetics exhibit the effect of converse magnetostriction.

4.1. *Magnetostrictive Relationships*

A current i amperes circulating in a coil of N turns produces in a closed ferromagnetic circuit a magnetic field of intensity, measured in ampere turns/m,

$$H = kNi/l$$

where l is the length of the magnetic circuit and k is a constant whose value depends upon its configuration. This field produces in the circuit a magnetic flux density $B = \mu H$, where μ is the permeability. The resulting strain S is proportional to the square of this flux density.

$$S = KB^2 \qquad (5.9)$$

where K is a material constant measured in m^4/Wb^2 which is positive for a material which expands upon magnetization and negative for one that contracts.

Owing to the polarization requirement, the flux density has two components B_0 and B_e, the polarizing and alternating components respectively.

Equation (5.9) therefore contains an alternating component

$$S = 2 KB_0B_e$$

or

$$S = \beta B_e \qquad (5.10)$$

where $\beta = 2 KB_0$ is the *magnetostrictive strain coefficient* defined as the strain produced per unit applied flux density and measured in m^2/Wb.

A tensile stress T, measured in newtons/m^2, also applied to the specimen results in a total strain

$$S = Ts + \beta\mu H_e \qquad (5.11)$$

where s is the elastic constant.

When a mechanical stress T is applied to a magnetostrictive element the resulting magnetic flux density is

$$B_e = \mu\beta T \qquad (5.12)$$

which leads to the alternative definition for β, namely the magnetic field intensity produced per unit applied stress measured in ampere turns/newton/m. If an external magnetic field is also applied, the flux density is increased to

$$B_e = \mu H_e + \mu\beta T \qquad (5.13)$$

The specimen can be operated under clamped conditions by

applying a stress which produces a strain equal and opposite to that produced by the magnetic field. S in eqn. (5.11) is zero and

$$T = \beta\mu H_e/s \quad \text{or} \quad T = \Lambda B_e \quad (5.14)$$

where $\Lambda = \beta/s$ is the *magnetostrictive stress coefficient* defined as the stress produced per unit flux density, measured in newtons/Wb.

Equations (5.10), (5.11), (5.12), (5.13) and (5.14) are magnetostrictive analogies of eqns. (5.3), (5.4), (5.1), (5.2) and (5.7) respectively. From these, analogies between various magnetostrictive and piezoelectric coefficients can be drawn, as shown in Table 5.2.

TABLE 5.2. MAGNETOSTRICTIVE AND PIEZOELECTRIC
COEFFICIENT ANALOGUES

Magnetostrictive coefficient	H	B	μ	$\mu\beta$	$\mu\Lambda$
Piezoelectric coefficient	E	D	ε	d	e

One major difference between these two sets of coefficients must be taken into account. The value of μ depends upon the polarizing field intensity H_0, though not linearly, and therefore B_e, $\mu\beta$ and $\mu\Lambda$ have to be specified for a particular point on the magnetization $(B-H)$ curve of the transducer material.

4.2. Magnetostrictive Vibrators

The basic features of three types of magnetostrictive transducer are shown in Fig. 5.5. Each consists of a magnetic circuit or core, energized by a coil carrying an alternating current, Polarization can be achieved by any of the methods listed in Section 4.1.

The material of the core has a relatively high electrical conductivity and the rapidly alternating magnetic field producing vibrations of the transducer also induces large currents called *eddy* currents in the core itself. These eddy currents dissipate, as heat, some of the electrical energy supplied to the transducer, so reducing its overall efficiency. Eddy current losses can be reduced by building up the magnetic circuit from a large number of laminations of thickness between 10^{-4} and 2×10^{-4} m, each lamination being insulated electrically from those adjacent to it.

The direction of the magnetic field applied to the core changes at twice the frequency of the applied electrical signal. Energy must be extracted from the electrical signal to realign the domains of the core material with each change of direction, thus reducing

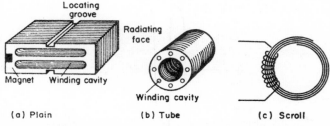

FIG. 5.5. Some typical magnetostrictive transducer configurations.

the efficiency of transduction once again. The actual loss incurred during each magnetic cycle is proportional to the area of the curve relating the established magnetic flux density B with the exciting magnetic field intensity H. A typical curve of this type is shown in Fig. 5.6 and is known as a *hysteresis loop*, the losses associated with it being called *hysteresis losses*. Materials for use in magnetostrictive transducers should therefore possess hysteresis loops with small areas.

Annealed-nickel is the most commonly employed material for magnetostrictive transducers. The annealing process produces an oxide layer on the surface of each lamination, so providing electrical insulation; it also reduces the area of the hysteresis loop.

Ferrites, which are ceramic-based oxides of solutions of iron in nickel, zinc or lead are also sometimes employed. They have magnetostrictive coefficients comparable with those of nickel and have the advantage of low electrical conductivity; they do not therefore have to be laminated. They are inferior mechanically and for this reason are best employed as hydrophones. If used as transmitters the large resulting strain would cause them to shatter. This can be alleviated to some extent, however, by pre-stressing them, so that they are always operated under conditions of compression.

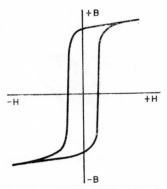

FIG. 5.6. A typical hysteresis ($B–H$) loop for annealed nickel.

5. Dynamic Properties of Transducers

Transducer action has previously been described in terms of the stress and strain resulting from the application of an electric or magnetic field. Now it is proposed to relate these quantities, which describe what takes place inside the transducer, to their corresponding external physical manifestations, force, pressure, particle-velocity, voltage and current. Magnetostrictive transducers will be treated separately on account of their greater complexity.

The following analysis strictly applies only to piezoelectric transducers but, provided account is taken of the fact that all electrostrictive constants have to be specified for a particular value of the polarizing electric field, it can be extended to electro-strictive transducers also.

5.1. *Dynamic Properties of Piezoelectric and Electrostrictive Transducers*

When a piezoelectric or electrostrictive element of area A and thickness l is used as a receiver, incoming acoustic waves cause vibrations of the element characterized by the velocity u at the surface.

From the definition of strain:

$$S = \frac{\text{extension}}{\text{original length}} = \int \frac{u\,\mathrm{d}t}{l}$$

it follows that

$$u = l\frac{\mathrm{d}S}{\mathrm{d}t} \tag{5.15}$$

The current flowing in any external circuit connected to the element is

$$i = A\frac{\mathrm{d}P}{\mathrm{d}t} \tag{5.16}$$

where P is the charge density induced per unit area.

Differentiating eqn. (5.8) gives

$$\frac{\mathrm{d}P}{\mathrm{d}t} = e\frac{\mathrm{d}S}{\mathrm{d}t}$$

therefore eqn. (5.16) becomes

$$i = Ae\frac{\mathrm{d}S}{\mathrm{d}t} \tag{5.17}$$

Equations (5.15) and (5.17) show that the current in the external circuit is proportional to the surface velocity of the transducer:

$$i = (Ae/l)\,u \tag{5.18}$$

Previously, in Section 5.1 of Chapter 3, the analogy between particle velocity and electric current has been drawn and Ae/l represents the *transformation ratio* α measured in coulombs/m, between the two analogous quantities

$$i = \alpha u \qquad (5.19)$$

There will be several values of α depending upon the choice of stress and electric field axes, the particular transducer material and in the case of electrostrictive transducers, the polarization axis. In general, eqn. (5.19) must be written

$$i_i = \alpha_{ih} u_h \qquad (5.19a)$$

A voltage V, applied across the element, used as a transmitter, produces in it an electric field of intensity

$$E = V/l$$

and a corresponding strain according to eqn. (5.7)

$$T = eV/l \qquad (5.20)$$

Stress is defined as force per unit area

$$T = F/A \qquad (5.21)$$

therefore eqn. (5.20) shows that a force

$$F = (Ae/l) V \qquad (5.22)$$

or

$$F = \alpha V \qquad (5.22a)$$

is produced as a result of the applied voltage. Again, force and voltage are analogous quantities related by the transformation ratio α. The general form of eqn. (5.22) is

$$F_i = \alpha_{ih} V_h \qquad (5.22b)$$

(This force accelerates the active mass of the transducer element and excites acoustic waves in the medium.)

It is possible also to determine the approximate efficiency of the transducer as an electroacoustic energy convertor by forming the ratio of the energy stored mechanically in the vibrating element, which is subsequently radiated as acoustic energy, to the energy supplied by the electrical source and stored in the electrical capacitance of the element.

The energy stored mechanically in the element is one-half of the product of force and extension, i.e.

$$W_m = \tfrac{1}{2} FSl$$

Using eqns. (5.6), (5.21) and (5.22a) this equation can be re-written as

$$W_m = \frac{1}{2} \frac{a^2 V^2 s l}{A} \qquad (5.23)$$

The energy stored electrically in the element is

$$W_e = \tfrac{1}{2} CV^2$$

where $C = A\epsilon/l$ is the capacitance of the element, so that

$$W_e = \frac{1}{2} \frac{A\epsilon}{l} V^2 \qquad (5.24)$$

Using the transformation ratio $a = Ae/l$ the ratio

$$\frac{W_m}{W_e} = \frac{\tfrac{1}{2}(A^2 e^2 s l V^2/l^2 A)}{\tfrac{1}{2}(A\epsilon V^2/l)} = \frac{e^2 s}{\epsilon} = k_c^2 \qquad (5.25)$$

k_c is called the *electromechanical coupling coefficient* and is dimensionless, being expressed in terms of the permittivity, the elastic constant and the piezoelectric stress constant only. It is a measure of the efficiency of the transducer but does not give absolute efficiency because losses within the element and its mounting and the acoustic loading by the medium itself must be taken into account.

The value of k_c is dependent upon the relative directions of the applied electric field and the resulting strain and is defined by the same system of subscripts as the other piezoelectric constants. For transducers possessing axes of circular symmetry, e.g. discs, the x-, y-, z-axes system cannot be used and then a *planar electromechanical coupling coefficient* k_p is referred to.

Typical values of k_c for various piezoelectric and electrostrictive materials together with their piezoelectric strain coefficients and Curie temperatures is shown in Table 5.3.

TABLE 5.3. PROPERTIES OF PIEZOELECTRIC AND ELECTROSTRICTIVE
TRANSDUCER MATERIALS

Material	Electro-mechanical coupling coefficient k_c (%)	Piezoelectric strain coefficient (coulomb/newton $\times 10^{-12}$)	Curie temperature (°C)
Quartz			
(X-cut)	10	−2·3	550
Rochelle salt			
(45° X-cut)	54	275	24
Lithium sulphate			
(Y-cut)	35	16·2	75
ADP			
(45° Z-cut)	29	24	120
Barium titanate			
Bar	50		
Plate	40	60–190	120–140
Radial	20		
Lead zirconate–titanate			
PZT–4a bar	76	80–320	320–490
PZT–5a bar	68		

5.2. Dynamic Properties of Magnetostrictive Transducers

The considerations discussed above can be applied similarly to magnetostrictive elements. When the element is used as a receiver incoming acoustic waves cause the core material to vibrate, so changing the magnetic field intensity. The law of electromagnetic induction states that a voltage is produced in a coil of N turns, placed round the core, equal to the rate of change of magnetic flux linkage. Thus

$$V = k \frac{\mathrm{d}}{\mathrm{d}t} (BAN)$$

or

$$V = kAN \frac{dB}{dt} \qquad (5.26)$$

where A and N are constants; and k depends upon the configuration of the magnetic circuit, being unity for a uniform toroid.

Differentiating eqn. (5.10) gives

$$\frac{dB}{dt} = \frac{1}{\beta} \frac{dS}{dt}$$

(ignoring the subscript e which is not necessary at present) and on applying eqn. (5.15) this reduces to

$$\frac{dB}{dt} = \frac{1}{\beta} \frac{u}{l}$$

where l now represents the effective length of the magnetic circuit.

Equation (5.26) reduces therefore to

$$V = \frac{kAN}{\beta l} \cdot u \qquad (5.27)$$

The transducer can be used as a transmitter by passing a current i through its coil which then produces a magnetic field of intensity

$$H = kNi/l$$

where k has the same significance as in eqn. (5.26). The corresponding magnetic flux density in the core is

$$B = \mu H = \mu k N i/l$$

and the resulting stress is given by eqn. (5.12):

$$T = kNi/\beta l$$

The force on the radiating mass of the transducer is therefore

$$F = TA = (kNA/\beta l) i \qquad (5.28)$$

The current through the coil is related to the voltage across it by the inductive impedance $Z = j\omega L$ of the transducer,

$$V = iZ \qquad (5.29)$$

Substituting eqn. (5.29) in eqns. (5.27) and (5.28) gives equations identical to those obtained for piezoelectric and electrostrictive elements, eqns. (5.19) and (5.22a), i.e.

$$i = \alpha u$$

$$F = \alpha V$$

where the transformation ratio $\alpha = kNA/\beta lZ$ depends upon the configuration of the magnetic circuit and is complex.

The electromechanical coupling factor k_c can be obtained by substitution of the analogous quantities of Table 5.2 into eqn. (5.25), thus:

$$k_c^2 \text{ (magnetostrictive)} = \frac{e^2 s}{\epsilon} = \frac{\mu^2 \Lambda^2 s}{\mu} = \mu \Lambda^2 s = \frac{\mu \beta^2}{s}$$

Typical values for k_c are given in Table 5.4 together with other properties of magnetostrictive materials.

TABLE 5.4. PROPERTIES OF SOME MAGNETOSTRICTIVE TRANSDUCER MATERIALS

Material	Composition	Electro-mechanical coupling factor k_c (%)	Magnetostrictive stress coefficient (Λ newtons/Wb $\times 10^6$)	Curie temperature
Annealed nickel	Ni	31	−20	358
Permalloy	Ni 45%, Fe 55%	12	2·7	440
Alfer	Al 13%, Fe 87%	27	6·7	500
Ferroxcube 4A	Manganese zinc ferrite	3	−90	190

6. Equivalent Circuit of a Mechanical Oscillator

All the transducers considered so far consist of two elements.

(1) An electrical storage element, i.e. an inductance or capacitance, plus
(2) A mechanical oscillator.

This oscillator consists of a mass m supported by a spring of *compliance* C_m as shown in Fig. 5.7 (a). In practice the mass m

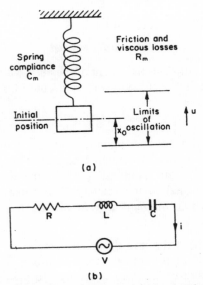

(a)

(b)

FIG. 5.7. (a) A mechanical oscillator and (b) its electrical equivalent.

is the vibrating mass of the transducer (of area A and thickness l) and the material of the transducer constitutes the spring. C_m is defined as the extension produced per unit applied force and is related to the constants of the transducer by

$$C_m = \frac{Sl}{F} = \frac{Sl}{TA} = \frac{sl}{A}$$

If the mass m is displaced a distance x the spring exerts a restoring force $-x/C_m$. If the mass is released, the spring moves the mass with an acceleration a, thus

$$ma = -x/C_m.$$

Rewriting this as a differential equation of motion in terms of the velocity u of the mass

$$m\frac{\mathrm{d}u}{\mathrm{d}t} + \frac{1}{C_m}\int u\mathrm{d}t = 0 \tag{5.30}$$

The solution to this equation is

$$u = \omega_0 x_0 \sin \omega_0 t$$

where x_0 is the initial displacement of the mass which subsequently executes indefinitely simple harmonic oscillations at a frequency $f_0 = \omega_0/2\pi$ where

$$\omega_0 = (mC_m)^{-1/2}$$

Loss of energy always occurs in real systems due to the presence of friction and viscous forces which cause the oscillatory motion to be damped. If the damping is small, the damping force is directly proportional to the velocity. This force can be written as $R_m u$, in which the *mechanical resistance R_m* has three components due to

(1) Losses within the transducer element.
(2) Losses within the transducer mounting.
(3) The loading of the transducer by the medium in contact with its radiating face or faces.

If losses within the transducer element and its mounting are assumed negligible (a reasonable approximation in a large number of cases) then the mechanical resistance is due solely to the loading imposed by the medium.

In a simple case when the transducer face is plane and large compared with the wavelength of the acoustic radiation, the radiated waves will be plane waves and the resistance presented to the transducer by the medium, termed the *radiation resistance*, R_r will be the product of the specific acoustic impedance per unit area and the radiating area, i.e.

$$R_m = R_r \rho c A$$

Equation (5.30) becomes

$$m \frac{\mathrm{d}u}{\mathrm{d}t} + R_m u + \frac{1}{C_m} \int u \mathrm{d}t = 0 \qquad (5.31)$$

If an external sinusoidal force $F(t)$ of angular frequency ω, of the form $F(t) = F\sin\omega t$ is applied to the transducer, the latter can be made to perform forced harmonic vibrations governed by the equation

$$F \sin \omega t = m \frac{\mathrm{d}u}{\mathrm{d}t} + R_m u + \frac{1}{C_m} \int u \mathrm{d}t$$

or

$$\omega F \cos \omega t = m \frac{\mathrm{d}^2 u}{\mathrm{d}t^2} + R_m \frac{\mathrm{d}u}{\mathrm{d}t} + \frac{u}{C_m} \qquad (5.32)$$

The solution of this equation under steady state conditions is

$$u = \frac{F}{R_m + \mathrm{j}(\omega m - 1/\omega C_m)} = \frac{F}{Z_m} \qquad (5.33)$$

where Z_m is the *mechanical impedance* of the transducer.

The differential equation governing the response of an electrical circuit, Fig. 5.7 (b), consisting of a series combination of inductance L, capacitance C, and resistance R and excited by a sinusoidal voltage at an angular frequency ω is

$$\omega V \cos \omega t = L \frac{\mathrm{d}^2 i}{\mathrm{d}t^2} + R \frac{\mathrm{d}i}{\mathrm{d}t} + \frac{i}{C} \qquad (5.34)$$

The solution of this equation under steady state conditions is

$$i = \frac{V}{R + \mathrm{j}(\omega L - 1/\omega C)} = \frac{V}{Z} \qquad (5.35)$$

Comparison of eqns. (5.32) and (5.33) with eqns. (5.34) and

(5.35) reveals that the analogy between electrical and mechanical quantities already observed, i.e. velocity–current and force–voltage, can be extended further to include: mass–inductance, compliance–capacitance and mechanical resistance–electrical resistance.

It is therefore possible to represent the mechanical oscillating element of a transducer by an equivalent electrical circuit which can then be combined with the electrical storage element to facilitate the analysis of the behaviour of the complete transducer when used with its associated electrical generating or receiving equipment.

Substituting eqn. (5.22a) into eqn. (5.33) gives a relationship between velocity and voltage

$$u = aV/Z_m \qquad (5.36)$$

and using this, eqn. (5.19) becomes

$$i = a^2 V/Z_m$$

or

$$Z = Z_m/a^2$$

Substituting for Z and Z_m

$$R + j\left(\omega L - \frac{1}{\omega C}\right) = \frac{R_m}{a^2} + j\left(\frac{\omega m}{a^2} - \frac{1}{a^2 C_m}\right)$$

from which can be obtained electrical elements equivalent to the mechanical oscillator

$$R = R_m/a^2$$

$$L = m/a^2 \qquad (5.37)$$

$$C = a^2 C_m$$

The equivalent electrical circuit can be expressed in either of the two forms of Fig. 5.8.

The mechanical oscillator can also be represented by a shunt combination of R, L and C using the equivalent elements of Fig. 5.9. It should be noted, however, that the elements R, L and C are not separately related to R_m, m and C_m, but each is a function of all three, being derived from the admittance relationsip $Y = \alpha^2/Z_m$. These equivalences apply only at the specified frequency; for different frequencies, different values of the elements are necessary.

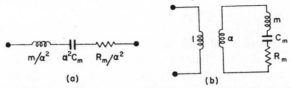

FIG. 5.8. Alternative series equivalent circuits for a mechanical oscillator.

FIG. 5.9. Shunt equivalent circuit of a mechanical oscillator.

7. Equivalent Circuits of Transducers

7.1. *Piezoelectric and Electrostrictive Transducers*

These both consist of a static capacitance C_0 and a mechanical oscillator. The motional current flowing in the oscillator is independent of the current flowing in the static capacitance and this leads to the shunt equivalent circuit shown in Fig. 5.10.

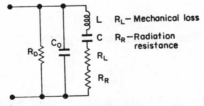

FIG. 5.10. Equivalent circuit of piezoelectric and electrostrictive transducers.

No capacitor is ever ideal and the resistance R_D is included to account for dielectric losses in the capacitive element.

For most underwater applications, transducers are operated at their fundamental resonance when

$$\omega_r = 1/(mC_m)^{1/2} = 1/(LC)^{1/2} \qquad (5.38)$$

and $Z_m = R_m$.

The element itself can be operated under two possible resonance conditions—*quarter-wave* and *half-wave resonance*.

Quarter-wave resonance is achieved by keeping one face of the element rigid while leaving the other free to vibrate. The thickness of the element is $\lambda'/4$ where λ' is the wavelength of sound within it. The one face is kept rigid by bonding it to a *backing plate* of a material with a higher characteristic acoustic impedance Z_b, also $\lambda/4$ thick. If the backing plate is also in contact along its opposite face with a medium of low acoustic impedance Z_n, e.g. air, it acts as a quarter-wave transformer and presents a very high impedance

$$Z = Z_b^2/Z_n \qquad (5.39)$$

to the element, as explained in Section 5.2 of Chapter 3, thereby keeping it rigidly clamped along one face.

With one face kept rigid the strain throughout the element can no longer be constant so the compliance C_m must be replaced by a *dynamic compliance* C_D and the effective mass by half the actual mass thus eqn. (5.38) becomes

$$\omega_r = (\tfrac{1}{2}mC_D)^{-1/2} \qquad (5.40)$$

The velocity of sound in the element is given by eqn. (3.6) as

$$c' = \sqrt{(\kappa/\rho)} = (s\rho)^{-1/2} = f_r'\lambda' \qquad (5.41)$$

where $\lambda' = 4l$. The resonant frequency f_r is

$$f_r = \frac{1}{(s\rho)^{1/2}} \cdot \frac{1}{4l}$$

which, from eqn. (5.40), also equals

$$1/[\pi(mC_D)^{1/2}]$$

Substituting $m = \rho Al$ and eqn. (5.30), gives

$$C_D = 8C_m/\pi^2 \qquad (5.42)$$

and this value of C_D must be used in the equivalent circuit in place of C_m.

Half-wave resonant operation can be obtained by bonding two quarter-wave elements back to back. The resulting half-wave element can be operated in one of two ways:

(a) With one face in contact with water, the other in contact with air giving what is known as *air-backed* resonance, or
(b) With both faces in contact with water giving *symmetrically-loaded* resonance.

In air-backed resonance the interface between the two halves is rigid but during symmetrically loaded resonance it acts as a velocity node. In both cases each quarter-wave section is in parallel with the other. As the effective length is half the true length, the dynamic compliance is

$$C_D = 4C_m/\pi^2 \qquad (5.42a)$$

for each parallel section, and the active mass is half the actual mass. A symmetrically loaded transducer having two faces in contact with the medium has two radiating surfaces each with a radiation resistance $R_R = \rho_c A$. Furthermore the voltage V applied to each half of the element is half the total voltage so the velocity of the radiating faces, as given by eqn. (5.36) is

$$u = aV/R_R \qquad (5.43)$$

An air-backed transducer has only one radiating face with a radiation resistance $R_R = \rho c A$, but the surface in contact with air is a soft boundary (see Chapter 4, Section 2.1) and produces almost total reflection of waves arriving from the radiating face. These reflected waves arrive back at the radiating face in phase with the waves generated there, so doubling the velocity of this face. Thus

$$u = 2aV/R_R \qquad (5.43a)$$

The power radiated by the transducer can be expressed in mechanical terms

$$W_m = u^2 R_R \qquad (5.44)$$

or in electrical terms

$$W_e = i^2 R = V^2/R \qquad (5.45)$$

For an air-backed transducer it follows from eqn. (5.43a) that eqn. (5.44) must be written as

$$Wm = 4a^2 V^2/R_R$$

Equation (5.45), however gives the power radiated

$$W = V^2/R$$

Comparing this with eqn. (5.44) gives

$$R = R_R/4a^2 \text{ (air-backed).}$$

For a symmetrically-loaded transducer eqns. (5.43) and (5.44), give

$$Wm = 2a^2v^2/R_R$$

but the power radiated is again

$$W = V^2/R$$

Comparing this with eqn. (5.44) gives

$$R = R_R/2\alpha^2 \text{ (symmetrically-loaded).}$$

In practice most transducers are air-backed to take advantage of the doubling of particle velocity and acoustic pressure so afforded.

Table 5.5 summarizes the basic relationships for the various modes of operation.

7.2 Magnetostrictive Transducers

These consist of a static or clamped inductance L_0 and a mechanical oscillator. Various equivalent circuits [2] are used to represent them but these frequently involve frequency-dependent elements in the branch representing the mechanical oscillator or

TABLE 5.5. BASIC RELATIONSHIPS FOR QUARTER- AND HALF-WAVE
RESONANT TRANSDUCERS

	Symbol	Unit	Quarter-wave resonance	Half-wave resonance	
				Air-backed	Symmetrically-loaded
Radiating surface area	A	m²	A	A	$2A$
Radiation resistance	R_R	kg/sec	$\rho c A$	$\rho c A$	$2\rho c A$
Transformation ratio	α	coulomb/m	Ae/l	$2Ae/l$	$2Ae/l$
Particle velocity	u	m/sec	$\alpha V/R_R$	$2\alpha V/R_R$	$\alpha V/R_R$
Acoustic power radiated	W	watts	$\alpha^2 V^2/R_R$	$4\alpha^2 V^2/R_R$	$2\alpha^2 V^2/R_R$
Motional resistance	R	ohms	R_R/α^2	$R_R/4\alpha^2$	$R_R/2\alpha^2$
Motional inductance	L	henrys	$m/2\alpha^2$	$m'/8\alpha^2$	$m'/8\alpha^2$
Motional capacitance	C	farads	$8\alpha^2 sl/\pi^2 A$	$8\alpha^2 sl'/\pi^2 A$	$8\alpha^2 sl'/\pi^2 A$

(N.B.: $l' = 2l$ and $m' = 2m$.)

complex transformation ratios relating mechanical and electrical circuits. The circuit of Fig. 5.11 is commonly used as it possesses neither of these disadvantages. The radiation load R_R appears in series with the mechanical oscillator Z_m. Additional elements can be added to take account of the complex nature of the eddy current loss in the core.

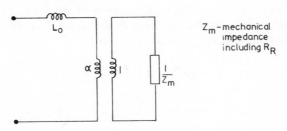

FIG. 5.11. Equivalent circuit of magnetostrictive transducers.

8. Frequency Response of Transducers

If the resistive elements are omitted from the equivalent circuits of Figs. 5.10 and 5.11 it is possible using techniques described in any textbook[3] dealing with elementary reactive network analysis to obtain the general form of the frequency response of the different types of transducer, as shown in Fig. 5.12.

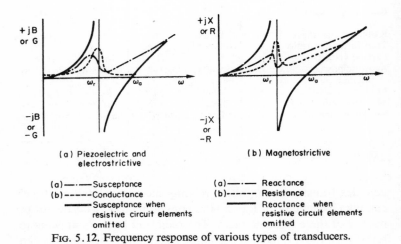

FIG. 5.12. Frequency response of various types of transducers.

The electrical admittance of the circuit representing piezo-electric and electrostrictive transducers is of the form $Y=G+jB$ where G is the conductive and B the susceptive component respectively and in this instance is given by

$$Y = j\omega C_0 - j\left(\frac{1}{\omega L - 1/\omega C}\right)$$
$$= j\omega \left(\frac{\omega^2 LCC_0 - C_0 - C}{\omega^2 LC - 1}\right)$$

which is wholly susceptive.

This admittance is zero when

$$\omega = 0$$

and

$$\omega = \omega_a = \sqrt{\left(\frac{C_0 + C}{LC_0C}\right)}$$

ω_a being the antiresonant angular frequency.

It is infinity when

$$\omega = \infty$$

and

$$\omega = \omega_r = \sqrt{(1/LC)},$$

ω_r being the resonant frequency.

The frequency dependence of the susceptance is shown by the full line of Fig. 5.12a. Reintroducing the resistors representing radiation resistance and losses gives a more realistic picture of the frequency response as represented by the dotted curve, *a*. In this response the resonance and antiresonance occur at slightly different frequencies. The dotted curve *b*, shows the dependence of the conductive component of admittance upon frequency under these conditions.

Similar arguments can be applied to the magnetostrictive equivalent circuit to give resonance and antiresonance near the frequencies

$$\omega_r = \sqrt{\left(\frac{1}{LC}\right)}$$

$$\omega_a = \sqrt{\left(\frac{L_0 + L}{LL_0C}\right)}$$

As will be shown in Section 5.9 it is common practice to tune out the clamped inductance or capacitance of a transducer using series capacitors or shunt inductors respectively. The frequency response then reduces to that of the mechanical oscillators alone shown in Fig. 5.13 and it is usual to match the input or output impedances of electrical receiving amplifiers and generators to the electrical resistance associated with the maxima of $|Z|$.

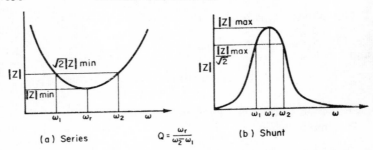

FIG. 5.13. Frequency responses of series and shunt mechanical oscillator equivalent circuits.

9. The Q-Factor of a Transducer

Many underwater applications, particularly those involving acoustic pulses of short duration or F.M. transmissions require acoustic transducers with as wide a relative frequency bandwidth as possible, i.e. with low Q-factors.

A transducer possesses two Q-factors, one the mechanical, Q_m, and one the electrical, Q_e, which are of course related. Q_m is defined as

$$Q_m = \omega_r m / R_m \qquad (5.46)$$

and depends upon the mode of operation of the transducer. Thus, for half-wave resonance, when $\omega_r = 2\pi c / 2l$ and $m = \frac{1}{2}\rho Al$, eqn. (5.46) gives the following values of Q_m for a plane transducer:

For air-backed operation

$$Q_m = \frac{\pi}{2} \frac{\rho c}{\rho_w c_w} \qquad (5.46a)$$

For symmetrically-loaded operation

$$Q_m = \frac{\pi}{4} \frac{\rho c}{\rho_w c_w} \qquad (5.46b)$$

For operation into water, with a different backing material

$$Q_m = \frac{\pi}{2} \frac{\rho c}{\rho_w c_w + \rho_b c_b} \qquad (5.46c)$$

Here ρc refers to the transducer, $\rho_w c_w$ to the water load and $\rho_b c_b$ to the backing material.

For an air-backed quartz transducer a Q-factor of about 15 is obtained if it is radiating into water but this is increased to about 50,000 if air is substituted as the load. For an air-backed barium titanate or nickel transducer a typical Q-factor of 200 in air is reduced to about 28 in water. In practice losses within the transducer mounting reduces these values of Q.

If a particularly low mechanical Q-factor is required, eqn. (5.46c) shows that the backing material should not be air but a lossy one of high characteristic acoustic impedance, such as steel or lead. If the backing is then made $\lambda/4$ thick the acoustic transformer so formed imposes a very high impedance given by eqn. (5.39) and the lowest possible Q-factor is obtained.

The electrical Q-factor is dependent only on the clamped capacitance or inductance and the electrical resistance presented at resonance. Thus for a piezoelectric or electrostrictive transducer

$$Q_e = \omega_r C_0 R$$

Substituting for the static capacitance $C_0 = A\epsilon/l$ and the radiation resistance $R_m = \rho c A$, and using eqns. (5.18), (5.25), (5.37) and (5.41) gives for a symmetrically-loaded element

$$Q_e = \frac{\pi^2}{2k_c^2} \bigg/ Q_m \qquad (5.47)$$

(A similar expression can be obtained from a consideration of magnetostrictive transducer action.) This expression shows that as the mechanical Q-factor is decreased by loading, the electrical Q-factor is increased. If a good transient or pulse response is required the electrical Q-factor presented to an electrical generator or receiving amplifier must be low. To satisfy this requirement also, the loading effect of the clamped electrical component C_0 or L_0 must be counteracted by the process of tuning out mentioned previously, and then damping the tuned circuit so formed by an external resistor. Thus the matching circuit for operating a piezoelectric or electrostrictive transducer would be of the form shown in Fig. 5.14. If low-Q operation is required, however, lower overall electroacoustic conversion efficiency must be expected.

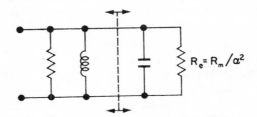

FIG. 5.14. Circuit for low-Q matching.

10. The Efficiency of Transducers

The absolute efficiency of a particular mounted transducer cannot be calculated owing to the difficulty of determining its internal losses and the losses associated with its mounting.

It could be determined empirically by measuring the acoustic intensity at a point remote from the transducer for a given electrical power input but this requires the use of a calibrated hydrophone which may not be readily available.

Efficiency is usually computed by first measuring the change of electrical admittance or impedance presented by the transducer as the frequency is varied in the region of its fundamental resonance, using a suitable impedance bridge or Q-meter.

10.1. *Piezoelectric and Electrostrictive Transducers*

The shunt equivalent circuit representations of these transducers necessitates their analysis in terms of their admittance $Y = G + jB$ where G is the conductance and B the susceptance. If G and B are measured using the circuit of Fig. 5.15 and a graph of G against B is plotted for different frequencies near the mechanical resonance a *motional admittance loop* of the type

shown in Fig. 5.16 (a), will be obtained. This will approximate
to a perfect circle provided the characteristic acoustic impedance
of the transducer element is greater than that of the medium into
which it is radiating and for this reason it is sometimes referred
to as a *circle diagram*. Certain precautions must be observed if

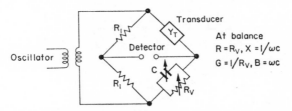

FIG. 5.15. Circuit for obtaining circle diagrams of piezoelectric
and electrostrictive transducers.

good circles are to be obtained. It is important to line the walls
of the containing vessel with absorbent material and so shape
them to prevent reflections which would otherwise form standing
waves.

The diameter of the circle diagram is inversely proportional to
the characteristic impedance of the loading medium and this fact
enables the efficiency of a transducer to be estimated. The
diameter (d_1) of the circle diagram obtained when the transducer
radiates into air (when the radiation resistance R_R of Fig. 5.10 is
effectively zero) gives directly the resistance R_L associated with
the internal and mounting losses only:

$$d_1 = \frac{1}{R_L} \qquad (5.48)$$

The diameter (d_2) of the circle obtained under normal operating
conditions gives the radiation resistance plus the loss resistance

$$d_2 = \frac{1}{R_R + R_L} \qquad (5.49)$$

These circles are shown schematically in Fig. 5.16 (b). In this

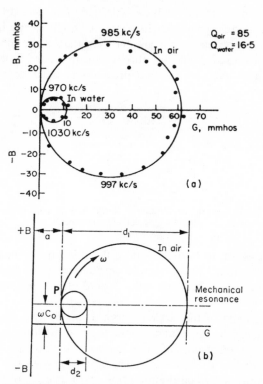

FIG. 5.16. (a) Typical circle diagram for an electrostrictive transducer. (The susceptive and conductive components due to the static capacitance and the dielectric loss respectively are not included.) (b) Schematic circle diagram for piezoelectric and electrostrictive transducers.

diagram the point P represents the static admittance of the element when it is effectively clamped mechanically and its motional admittance zero; this occurs when the element is operated at an even harmonic of the fundamental resonance. The distance a therefore gives the dielectric loss resistance R_D, since

$$a = 1/R_D \qquad (5.50)$$

The efficiency of a transducer is defined as

$$\eta = \frac{\text{power radiated into the load}}{\text{total power input}} \qquad (5.51a)$$

$$= \frac{P_R}{P_R + P_L + P_D} \qquad (5.51b)$$

If the electric potential difference applied to the circuit of Fig. 5.10 is a sinusoid of peak value V, at the frequency of resonance

$$\text{the radiated power } P_R = \frac{V^2 R_R}{2(R_L + R_R)^2}$$

$$\text{the mechanical loss } P_L = \frac{V^2 R_L}{2(R_R + R_L)^2}$$

$$\text{and the dielectric loss } P_D = \frac{V^2}{2R_D}$$

Equation (5.51b) can therefore be rewritten as

$$\eta = \frac{R_R R_D}{(R_R + R_L)(R_R + R_L + R_D)}$$

and can be expressed in terms of the circle diagram Fig. 5.16 (b) using eqns. (5.48), (5.49) and (5.50) as

$$\eta = \frac{d_2(d_1 - d_2)}{d_1(a + d_2)} \qquad (5.52)$$

10.2. *Magnetostrictive Transducers*

This type of transducer is represented by a series equivalent circuit so the circle diagram is obtained in terms of its resistance R and its reactance X, as measured by the circuit of Fig. 5.17. A typical circle diagram is shown in Fig. 5.18 (a) together with schematic diagrams for operation into air and water loads, Fig. 5.18 (b). One major difference between these circles and corresponding circles obtained for piezoelectric and electrostrictive transducers is that they are inclined at an angle θ to the resistive axis due to the complex core impedance associated with the flow of residual eddy currents.

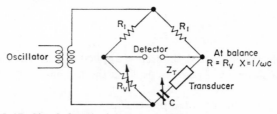

Fig. 5.17. Circuit for obtaining circle diagram of magnetostrictive transducers.

As in the case of piezoelectric and electrostrictive transducers, the various elements in the equivalent circuit of Fig. 5.11 can be expressed in terms of the circle diagrams, thus:

$$d_1 \cos \theta = R_L \qquad (5.53)$$

$$d_2 \cos \theta = \frac{R_R R_L}{R_R + R_L} \qquad (5.54)$$

and

$$a = R_C \qquad (5.55)$$

The efficiency of a transducer is defined by eqn. (5.51) and can be expressed in the form

$$\eta = \frac{P_R}{2IV \cos \theta} \qquad (5.56)$$

since I and V are both peak values of a sinusoid where the radiated power

$$P_R = \frac{I^2}{2} \left(\frac{R_L}{R_R + R_L} \right)^2 R_R$$

and I and V are the motional current through the transducer and the applied voltage respectively.

At resonance

$$V = I \left(R_c + \frac{R_R R_L}{R_R + R_L} \right)$$

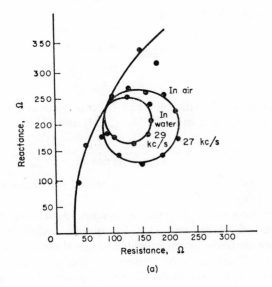

(a)

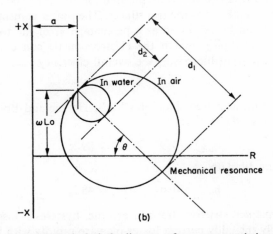

(b)

FIG. 5.18. (a) Typical circle diagram of a magnetostrictive transducer. (b) Simplified schematic circle diagram of magnetostrictive
transducers.

therefore eqn. (5.56) becomes

$$\eta = \frac{R_x^2}{R_R(R_c + R_x) \cos \theta}$$

where

$$R_x = \frac{R_R R_L}{R_R + R_L}$$

This expression for efficiency can be expressed in terms of the circle diagram using eqns. (5.53), (5.54) and (5.55), thus:

$$\eta = \frac{d_2(d_1 - d_2)}{d_1(a + d_2 \cos \theta)} \tag{5.57}$$

which is similar to the expression obtained for piezoelectric and electrostrictive transducers.

For properly mounted transducers eqns. (5.52) and (5.57) give efficiencies of 99, 90 and 75% for quartz, nickel and barium titanate respectively, when operated in the thickness mode.

The overall efficiency of conversion from electrical a.c. input power to acoustically radiated power depends also upon the amount of power lost in the circuit used for matching the electrical generator or receiver to the transducer. The power transmitted by this circuit can be expressed as a percentage η_1 of the total electrical power supplied and if the electroacoustic conversion efficiency of the transducer is η_2 the overall efficiency

$$\eta_{\text{total}} = \eta_1 \eta_2$$

For operation at a frequency of 1 Mc/s, Heuter and Bolt[4] give the following efficiencies:
For quartz

$$\eta_{\text{tot.}} = 0 \cdot 9 \times 0 \cdot 99 = 89\%$$

For barium titanate

$$\eta_{\text{tot.}} = 0 \cdot 6 \times 0 \cdot 8 = 48\%$$

In a magnetostrictive transducer, the hysteresis losses and particularly the eddy current losses increase rapidly with increase in operating frequency and therefore its electroacoustic conversion efficiency decreases. Heuter and Bolt give typical conversion efficiencies of 74% at 15 kc/s falling to only 0·08% at 500 kc/s.

11. Practical Aspects of Transducer Mountings

Mechanical vibrations of a transducer element are readily coupled into a water load so no special techniques need be employed to improve efficiency; this greatly simplifies the design of transducer housings. Sea-water must not be allowed access to a piezoelectric or electrostrictive element for two reasons: firstly, it would corrode the electrodes, and secondly, it would provide an alternative path for current flow between the electrodes, so shorting out the element.

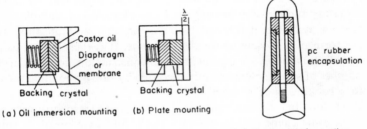

(a) Oil immersion mounting   (b) Plate mounting

(c) Encapsulated mounting

FIG. 5.19. Some typical transducer mountings.

The coil of a magnetostrictive transducer will generally be provided with heavy insulation, so strictly this type needs no further protection; but it is often provided with a housing which is designed to prevent marine organisms fouling its radiating face.

Specially shaped housings are usually required for all transducers mounted on the hull of a ship to streamline the flow of water past them. This streamlining reduces the noise induced in a transducer as a result of turbulent water rushing past it and prevents the formation of an obscuring bubble stream.

All of these requirements can be provided by the mountings

shown diagrammatically in Fig. 5.19 in terms of piezoelectric or electrostrictive transducers. In the first of these, the element with its chosen backing is mounted in an insulating liquid, usually castor oil, and coupled into the water by means of a thin metallic diaphragm, e.g. aluminium less than $\lambda/30$ in thickness, or by a "ρc-rubber" membrane whose characteristic acoustic impedance matches that of sea-water. The second method in which the element is bonded to a metal plate, $\lambda/2$ in thickness, affords far greater mechanical protection, but the mechanical Q-factor of the complete transducer is increased considerably. The encapsulation technique is rapidly gaining in popularity and is ideally suited to the construction of sensitive hydrophones and transmitters not subjected to great mechanical stresses.

Where crystal or ceramic transducers with a particularly large radiating surface area are required and elements of suitable size are not available, it is usual to build the required transducer in the form of a mosaic of smaller elements. It is then known as a *mosaic transducer*. A typical mosaic construction consists of a number of similar elements mounted side by side by bonding them to a backing plate of steel.

At low frequencies, quarter- or half-wave resonant crystal or ceramic elements become very thick, and as a result of this they are difficult to manufacture and require excessive driving voltages. It is then desirable to operate a thin element in such a way that it resonates at a frequency lower than that characterized by its thickness. This is accomplished by bonding metal blocks to its opposite faces to form a sandwich type of construction. Some typical examples of such *sandwich transducers* are shown in Fig. 5.20.

The simplest form of sandwich transducer consists of a thin ceramic element to which is bonded a block of metal $\lambda/4$ in thickness at the required frequency of operation, which acts as an acoustic transformer (see Section 7.1) and imposes only a resistive load on the element. The element itself will be operated far below its natural resonance and its static capacitance will make it appear reactive. If an air-backed block of metal less than $\lambda/4$ in thickness is bonded to its other face the loading it imposes on the element appears mass-like, i.e. inductive. If the thickness d of the ceramic element and the backing block are chosen in accordance with the expression given by Heuter and Bolt[4]

$$\tan \frac{2\pi d_e}{\lambda_e} \tan \frac{2\pi d_b}{\lambda_b} = \frac{\rho_e c_e}{\rho_b c_b} \qquad (5.58)$$

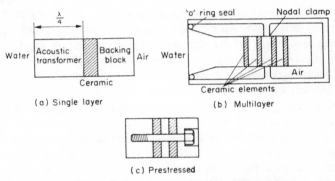

FIG. 5.20. Typical sandwich transducer constructions.

where the indices e and b refer to the element and back plate respectively, the inductance of the back plate will resonate with the capacitance of the element at the design frequency and the transducer will present a purely resistive load to its associated electrical equipment.

More complex forms of sandwich construction are often used; for example, in some the area of the radiating face is different from that of the rest of the sandwich and a conical matching section has then to be employed. Other designs use a multilayer sandwich construction in which many ceramic elements are used to increase the acoustic power radiated. In very high power transmitters there is the danger that the resulting large amplitude vibrations may shatter the element or its bond to the metal loading sections. To overcome this a bolt is passed along the axis of the transducer which is used to prestress it so that all elements and joints will always operate under compression. Details of pre-stressed multilayer transducers are given by Schofield,[5] and Heuter and Bolt describe many other sandwich configurations.

12. High Intensity Transmitters

The long range geological investigations mentioned in Chapter 1 require extremely intense sound sources and for this reason explosive sources are used and charges of up to 150 kg of TNT may be detonated. An explosive source has the disadvantages of being dangerous to handle and expensive; and it produces a non-repetitive waveform. For shorter range applications such as sub-bottom profiling it is preferable to use an electroacoustic transducer of the *sparker* or *boomer* type. Both of these devices derive their power from the energy stored in a number of large charged capacitors.

For the sparker the capacitors are charged to a high potential, typically 25 kV, and then discharged via a triggered gas discharge tube through a pair of electrodes immersed in the sea. This discharge vaporizes the water between the electrodes and produces intense acoustic radiation which has a very wide frequency spectrum.

The boomer is an electromechanical device and consists of a flat spiral coil held lightly in contact with an aluminium plate by a spring. When the capacitor bank is discharged through the coil an intense magnetic field is set up which induces eddy currents to flow in the aluminium plate. These eddy currents in turn produce another magnetic field which interacts with the primary field and forces the coil and the plate apart.

This produces an almost unidirectional pressure pulse which has a wide frequency spectrum. In the case of a boomer $0 \cdot 25$ m² in area energized by a 1000 joule power unit the duration of the pulse is of the order of 1 msec, and produces an acoustic pressure wave equivalent to that produced by the detonation of 1 g of TNT.

References

1. MASON, W. P., *Piezoelectric Crystals and their Applications to Ultrasonics*, Van Nostrand, New York (1950).
2. FISCHER, F. A., *Fundamentals of Electroacoustics*, Interscience, New York (1953).
3. TUCKER, D. G., *Elementary Electrical Network Theory*, Pergamon, London (1964).
4. HEUTER, T. F., and BOLT, R. H., *Sonics*, Wiley, New York (1955).
5. SCHOFIELD, D., *Transducers, Underwater Accoustics*, V. M. Albers (Editor), Plenum Press, New York (1963).

CHAPTER 6

Directivity of Transducers and Arrays

1. Introduction

In the previous chapter the mechanisms and efficiency of transducer action have been studied, and some information given on the design of transducers and measurement of their properties. However, attention was concentrated on the conversion of electrical energy into acoustic energy, and vice versa, and nothing was said about the directional properties of the sound radiation and reception. This latter aspect is the subject of the present chapter.

It was seen in Chapter 1 that in the majority of applications of underwater acoustics there is a need for information on the direction of the object or sound source being detected, and sometimes very high precision (e.g. an angular error of a small fraction of a degree) is specified. The most obvious way to achieve high angular accuracy is to use a receiving transducer system which responds effectively over only a very small angle of the space around it. It is often also advantageous when transmission is involved as well as reception (as in sonar systems) to use a transmitting system which energizes (or "insonifies") only a very small angle. Such systems, which respond in transmission or reception over a restricted range of directions, are called *directional systems*. In transmission the sound is thought of as being confined to a *beam*; and by analogy receiving systems are also referred to as having a beam, which is the solid angle over which their response is effective. The property of confining response (whether on transmission or reception) to a limited angle is called *directionality* or *directivity*; the latter term has a more quantitative connotation than the former, but neither has (on its own) a quantitative definition.

Transducers of finite size, with the exception only of perfectly spherical transducers with an entirely radial surface motion, always have some measure of directionality. A point transducer (i.e. one of infinitesimal dimensions), and the spherical transducer with radial motion, respond equally in all directions if they are in a uniform medium of infinite extent and are called *omnidirectional*.

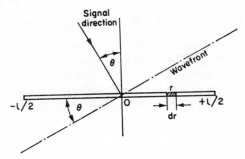

Fig. 6.1. Line array for reception.

A directional transducer system may comprise a single transducer with a radiating surface of shape designed to give the desired directionality, it may comprise an array of omnidirectional transducers so arranged as to give a particular directionality, or it may combine both these arrangements by using an array of directional transducers. When an array is used, particularly for reception (but also to a more limited extent on transmission), the individual transducers (or "elements") may be given different treatment in their electrical connections and so the idea of *signal processing* becomes important here. The whole subject is one of great complexity in which much research is still being done. Consequently it will be possible, in this book, to give merely some of the simpler and more basic ideas.

Throughout this chapter it will be assumed that far-field conditions apply, i.e. directional patterns are measured at a range very large compared with the dimensions of the array. At distances relatively close to the array, roughly within the region in which the width of the nominal beam (as determined from the angle of the far-field beam) is less than the length of the array, much more complex results are obtained. In view of their relatively small practical importance, results for the "near-field" (or Fresnel) region will not be discussed in this book.

2. Basic Fourier Transform Relationship between Excitation (or Sensitivity) Distribution and Directional Pattern

From the academic point of view, if not from the practical, the most fundamental relationship in considering directionality is the Fourier transform which relates the distribution of excitation (in a transmitter) or sensitivity (in a receiver) over the face of the transducer or array to the distribution of directional response in space.

2.1. *Linear Transducers and Arrays*

For simplicity, consider a transducer or array† which has length but no width or depth. Clearly, in any plane which is normal to the finite axis of the array, the array appears as a point transducer and is omnidirectional in this plane. But in any plane containing the finite axis, the response is directional. Consider any such plane, as shown in Fig. 6.1, and assume that the array is used for receiving an acoustic signal which comes from a distance long compared with the length l of the array, so that the received wave is a plane wave, with a wavefront making an angle θ to the axis of the array. Assume that the wave received at the centre of the array is $\cos \omega_q t$, with a normalized amplitude of unity. At a point on the array distant r from the centre, the wave is $\cos (\omega_q t - Kr)$, where $K = (2\pi/\lambda) \sin \theta$, and λ is the wavelength in the medium.

† In future we shall use the term "array" to mean "transducer or array".

Now the sensitivity of the array is not considered to be uniform along its length, but to have a value $T(r)$ at the point distant r from the centre. Thus over an infinitesimal element of length dr, as shaded in the diagram, the electrical response is $T(r) \cdot dr \cdot \cos(\omega_q t - Kr)$. The total electrical output from the array is therefore

$$\int_{-l/2}^{+l/2} T(r) \cdot \cos(\omega_q t - Kr) \cdot dr \tag{6.1}$$

Since the sensitivity of the array is zero for points outside its length, the limits of integration may be made infinite without altering the result, and we have therefore as the total output:

$$\int_{-\infty}^{\infty} T(r) \cdot \cos(\omega_q t - Kr) \cdot dr \tag{6.2}$$

$$= \int_{-\infty}^{\infty} T(r) \cdot (\cos \omega_q t \cdot \cos Kr + \sin \omega_q t \cdot \sin Kr) \cdot dr$$

$$\left. \begin{array}{l} = \left(\displaystyle\int_{-\infty}^{\infty} T(r) \cdot \cos Kr \cdot dr \right) \cos \omega_q t \\[3mm] + \left[\displaystyle\int_{-\infty}^{\infty} T(r) \cdot \sin Kr \cdot dr \right] \sin \omega_q t \end{array} \right\} \tag{6.3}$$

Now in eqn. (6.3) the first term in round brackets is recognized as the *cosine Fourier transform* of $T(r)$ and the second term in square brackets as the *sine Fourier transform*. Further information on Fourier transforms may be obtained from books on the subject.[1]

Most commonly one is not concerned with the phase of the resultant signal, and the directional information required is merely the relative amplitude response in different directions. This is obtained by taking the modulus of the two amplitude terms in eqn. (6.3), so that the *directional function*, which we may write $D(K)$, becomes

$$D(K) = \sqrt{\left[\left(\int_{-\infty}^{\infty} T(r) \cdot \cos Kr \cdot dr \right)^2 \right.}$$

$$\left. + \left(\int_{-\infty}^{\infty} T(r) \cdot \sin Kr \cdot dr \right)^2 \right] \quad (6.4)$$

This is clearly a function of K and expresses the total response of the array for any particular signal direction as represented by K. If a graph is plotted of the total amplitude response against θ, a *directional pattern* is obtained, and it would usually be drawn in

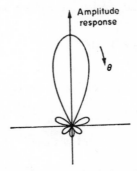

FIG. 6.2. Typical directional pattern in polar coordinates.

polar form as illustrated in Fig. 6.2. For practical purposes this is a very convenient form of the directional pattern. It can be measured by rotating the array while the signal direction is held constant. But for academic and design purposes it is more useful to plot the total response against K on rectangular coordinates, as illustrated in Fig. 6.3. An important advantage of the latter form is that K may take any value, including values for which sin θ exceeds unity and for which, therefore, θ has no real value. Obviously the polar plot in terms of θ is confined to θ within the range $-\pi$ to $+\pi$ rad, and although no physical meaning can be given to angles outside this range, yet a knowledge of the response outside this range can sometimes be useful, as will be seen in Section 4.4. Moreover, the form of Fig. 6.3 is more general, since it can be drawn before the length of the array is specified, and therefore before the number of lobes in Fig. 6.2 is known.

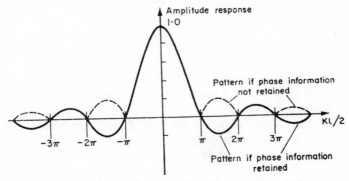

FIG. 6.3. Typical directional pattern, plotted against K (proportional to $\sin \theta$) on rectangular coordinates. N.B. This is, in fact, the curve of $(\sin Kl/2)/(Kl/2)$.

It will be observed from the directional patterns illustrated in Figs. 6.2 and 6.3 that there are typically several beams or *lobes* in a pattern. The *main lobe* is, of course, what is normally meant by the beam of the array, but the *minor* or *secondary lobes* are present in most patterns and give rise to a number of disadvantages including possible ambiguities in directional determination. The way these minor lobes arise will be clearer when some specific examples are taken.

Often, in practice, the array is *symmetrical* so that $T(r)$ is an even function, i.e. $T(r) = T(-r)$. Then clearly

$$\int_{-\infty}^{\infty} T(r) \cdot \sin Kr \cdot \mathrm{d}r = 0 \tag{6.5}$$

and

$$D(K) = \left| \int_{-\infty}^{\infty} T(r) \cdot \cos Kr \cdot \mathrm{d}r \right| \tag{6.6}$$

But we can also see that in this symmetrical case there are no phase variations with direction, other than complete reversals, since the coefficient of sin $\omega_q t$ in eqn. (6.3) is always zero; but complete phase reversals can occur when the cosine transform changes sign. If information on these is to be retained, we must write

$$D(K) = \int_{-\infty}^{\infty} T(r) \cdot \cos Kr \cdot dr \qquad (6.7)$$

Equation (6.7) is represented by the full lines in Fig. 6.3, and eqn. (6.6) by the dotted lines, the illustrative case taken being a symmetrical one.

If, in the general case, we do wish $D(K)$ to represent phase variations with direction, we must use the *complex Fourier transform*, which from eqn. (6.3) we see to be

$$D(K) = \int_{-\infty}^{\infty} T(r) \cdot \cos Kr \cdot dr - j\left(\int_{-\infty}^{\infty} T(r) \cdot \sin Kr \cdot dr \right) \qquad (6.8)$$

which may be written in exponential form as

$$D(K) = \int_{-\infty}^{\infty} T(r) \cdot \exp\left(-jKr\right) \cdot dr \qquad (6.9)$$

which is the best known and standard form.

The function $T(r)$ is often called the *taper function* of the array; the name probably derives from the fact that arrays are frequently used with $T(r)$ of symmetrical form, "tapering" from a maximum sensitivity at the centre to minimum sensitivity at the ends of the array, since this gives a reduction in the amplitudes of the minor lobes of the directional pattern.

It has been shown, then, that the directional function $D(K)$ is related to the sensitivity or taper function $T(r)$ by the Fourier transform. Correspondingly, the inverse transform gives the $T(r)$ required to produce a specified $D(K)$. In principle, therefore, this solves the design problem of an array. A suitable $D(K)$ is determined from operational considerations, and a $T(r)$ is calculated which will give it. Unfortunately, the integral will often be difficult to evaluate, and even more often the corresponding $T(r)$—supposing it can be calculated—will be impracticable or undesirable in other ways. A more practical approach to array design is given in Section 3. We will merely illustrate the Fourier transform method by some simple examples where it works well.

EXAMPLE 1. *Uniform line array*: $T(r) = 1/l$ from $r = -l/2$ to $r = l/2$, elsewhere it is zero. This arrangement is symmetrical so, using eqn. (6.6),

$$D(K) = \frac{1}{l} \cdot \int_{-l/2}^{l/2} \cos Kr \cdot dr = \frac{\sin (Kl/2)}{Kl/2} \qquad (6.10)$$

which is the well known curve usually referred to as the $(\sin x)/x$ curve; and it is, indeed, this curve which is used in Fig. 6.3. If the main lobe is of unit height, then the minor lobes have heights of approximately $0 \cdot 217$, $0 \cdot 128$, $0 \cdot 091$, $0 \cdot 071$, etc., taken in order from the origin. Note that the beamwidth is exactly inversely proportional to the length provided the scale of $\sin \theta$ is used to express it.

EXAMPLE 2. *Cosine taper array*: $T(r)=(2/l) \cos (\pi r/l)$ from $r = -l/2$ to $r = l/2$, elsewhere it is zero. This gives a taper from maximum at the centre of the array to zero at the ends. Again using eqn. (6.7),

$$D(K) = \frac{2}{l} \int_{-l/2}^{l/2} \cos \frac{\pi r}{l} \cdot \cos Kr \cdot dr$$

$$= \frac{1}{l} \int_{-l/2}^{l/2} \cos \left(K + \frac{\pi}{l}\right) r \cdot dr + \frac{1}{l} \int_{-l/2}^{l/2} \cos \left(K - \frac{\pi}{l}\right) r \cdot dr$$

$$= \frac{\sin (Kl/2 + \pi/2)}{Kl/2 + \pi/2} + \frac{\sin (Kl/2 - \pi/2)}{Kl/2 - \pi/2} \qquad (6.11)$$

This directional pattern is shown in Fig. 6.4, normalized to unity peak height by multiplying by a factor $\pi/4$. It will be seen that the minor lobes are greatly reduced in height, the largest being only 0·07 instead of the 0·217 of the previous pattern; but the *beamwidth* (i.e. the width of the main lobe between the points of half-power (i.e. 3 dB) response) has been increased by about 28%.

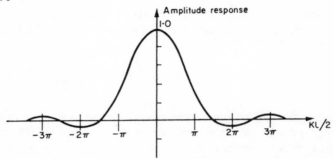

FIG. 6.4. Directional pattern for cosine taper.

EXAMPLE 3. *Linear point array*, consisting of n point transducers ($n = 2m + 1$, m even or odd integer or zero, n odd integer) uniformly spaced along a straight line, and each with unit sensitivity. Here we replace r by a number s identifying each point transducer, or "element", so that eqn. (6.7) becomes

$$D(K) = \sum_{s=-m}^{s=+m} \cos Ksd \qquad (6.12)$$

where $d = $ spacing between elements.

By a well known trigonometric transformation, this can be written as

$$D(K) = \frac{\sin (nKd/2)}{\sin (Kd/2)} \qquad (6.13)$$

which is a repetitive pattern, illustrated in Fig. 6.5 for $n = 9$, and usually known as a *diffraction pattern*. The pattern is plotted on an abscissa scale of $\sin \theta$ ($= \lambda K/2\pi$) because this makes the intervals between successive zeros equal to the reciprocal of the notional length of the array (i.e. *nd*) expressed in units of the wavelength. Although it has been assumed above that *n* is odd, a similar result is readily obtained if *n* is even.

A pattern of this kind can be very useful when the ambiguity due to the repetitive main lobes either does not matter, or does not exist because: (i) the medium is insonified only over the angular range of one particular period of the pattern; or (ii) the

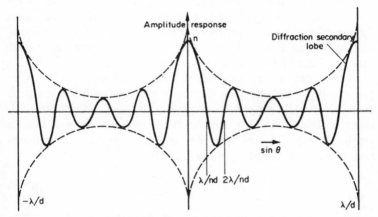

FIG. 6.5. Diffraction pattern of a linear point array.

spacing d/λ is such that only one period of the pattern occurs in the range of real angles, i.e. in the range of θ where $\sin \theta \leqslant 1$. At very low frequencies, where d would perhaps be many metres or even kilometres, a continuous array is not practicable, and a point array has of necessity to be used when a directional response is required.

Two extreme cases of the linear point array are of interest and importance. Firstly, take $n = 2$. This is the *interferometer*, so called because its response on transmission is the result of interference between two waves; it is well known in other applications of wave theory. Here we see that the directional pattern is merely

a cosinusoid:
$$D(K) = 2 \cos (Kd/2) \qquad (6.14)$$

The beamwidth of the lobes of this pattern is one-half of that of the main lobe of the uniform array (Fig. 6.3), assuming that the spacing d of the elements in the interferometer is equal to the length l of the uniform array; but, of course, there is the ambiguity arising from the repetition of equal lobes in the pattern to counteract this apparent advantage. Nevertheless, as shown in Chapter 2, Section 4, this objection tends to disappear when a wide band signal is used in place of the sine wave assumed here, because then the directional pattern approaches the $(\sin x)/x$ correlation function shown in Fig. 2.7, which is the same as the directional pattern of the uniform array for a sinewave signal.

As the other extreme case of the linear point array, we consider $n \to \infty$. If we assume that the array remains of finite length l, then $d \to 0$ while $nd = l$. Since we do not want an infinite peak value of $D(K)$, we must now assume that each element has an amplitude sensitivity of $1/n$. We then have, in effect, a uniform continuous array. Making these substitutions in eqn. (6.13),

$$D(K) = \frac{\sin (Kl/2)}{Kl/2} \qquad (6.15)$$

where, of course, we have put $\sin (Kd/2) \to Kd/2$ as $d \to 0$. This pattern is naturally identical to that obtained directly in eqn. (6.10).

EXAMPLE 4. *Linear array of elements each of finite length.* Let the elements each be of length a uniformly spaced at distance d between centres, and of sensitivity $1/a$. It can be assumed initially that $a < d$. Then, numbering the elements as in the previous example, and retaining r in its previous significance, as shown in Fig. 6.6 (a), we see that eqn. (6.7) becomes

$$D(K) = \sum_{s=-m}^{s=+m} \frac{1}{a} \int_{sd-a/2}^{sd+a/2} \cos Kr \, . \, dr$$

$$= \sum_{s=-m}^{s=+m} \cos Ksd \cdot \frac{\sin (Ka/2)}{Ka/2}$$

$$= \frac{\sin (Ka/2)}{Ka/2} \cdot \frac{\sin (nKd/2)}{\sin (Kd/2)} \qquad (6.16)$$

Thus the directional pattern is now the product of two patterns: the directional pattern of each element alone, and that of the corresponding linear point array. The second term is often called the *array factor*, and it is easily seen why.

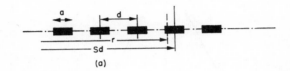

(a)

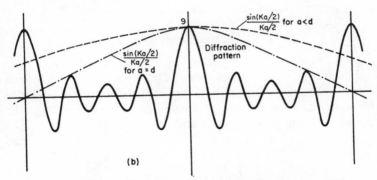

(b)

Fig. 6.6. (a) Part of an array of elements each of finite length. (b) Directional pattern for array shown in (a) when (i) $a < d$ and (ii) $a = d$.

When $a = d$, we have a continuous uniform array, and eqn. (6.16) becomes the same as eqns. (6.10) and (6.15) when the peak height is reduced to unity by dividing by n.

There is no reason in principle why a should not exceed d by overlapping the elements, and, indeed, this has been used in practice when special requirements have made it advantageous.

The significance of eqn. (6.16) can be seen more readily with the help of Fig. 6.6 (b) which is for $n = 9$. It can there be seen how when $a = d$ the zeros of the first factor exactly coincide with the main repetitive peaks of the array factor and so reduce the overall curve to the (sin x)/x form. This view of the array will be useful later, in Section 3.3, when we deal with electrical deflection of beams.

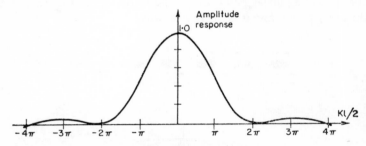

FIG. 6.7. Directional pattern for linear taper.

EXAMPLE 5. *Linear taper array*: $T(r) = [1 - 2 \mid r \mid /l]/l$, from $r = -l/2$ to $r = l/2$; elsewhere zero. This is somewhat similar to the cosine taper of Example 2, having maximum sensitivity in the centre, tapering to zero at the ends. So from eqn. (6.7),

$$D(K) = \frac{1}{l} \int_{-l/2}^{l/2} (1 - 2 \mid r \mid /l) \cos Kr \cdot dr$$

which can be shown to be

$$D(K) = \frac{1 - \cos (Kl/2)}{K^2 l^2/4} = \frac{1}{2} \left[\frac{\sin (Kl/4)}{Kl/4} \right]^2 \quad (6.17)$$

This is shown as a pattern in Fig. 6.7; although the general shape of the beam and minor lobes resembles that of the cosine taper (Fig. 6.4), there is the difference that now all minor lobes are of the same polarity as the main lobe. This is sometimes important.

Note that to obtain a normalized peak height of unity for this pattern, a normalizing factor of 2 is required.

2.2. *Planar Transducers and Arrays*

In practice, all transducers and arrays have finite dimensions on more than one axis, so that although the line array is very useful analytically, it is usually an idealization. Nevertheless, the work of the previous section is applicable to the most common kinds of transducer or array, namely, those in which the radiating or receiving elements or surfaces lie in a plane. In the case of a single transducer, this merely means that it is has a flat radiating surface.

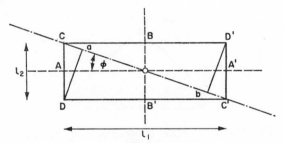

FIG. 6.8. Rectangular transducer.

2.2.1. *Rectangular Transducer*

We have to determine the directional pattern in any plane normal to the surface or plane of the array. Consider a rectangular transducer as shown in Fig. 6.8. It will be assumed that the sensitivity is uniform over the whole surface. If the directional pattern is required in a plane containing the long axis AA', we can regard the transducer merely as a line transducer of length l_1 and of uniform sensitivity along its length corresponding to the uniform width l_2 of the transducer relative to this axis. The directional pattern is thus

$$D_A(K) = \frac{\sin{(Kl_1/2)}}{Kl_1/2} \qquad (6.18a)$$

Similarly, the directional pattern in a plane containing the short axis BB' is

$$D_B(K) = \frac{\sin{(Kl_2/2)}}{Kl_2/2} \qquad (6.18b)$$

which obviously has a wider beamwidth but is otherwise of the same shape. But when we require the directional pattern for any other axis we find a different situation.

Suppose we want the directional pattern for the plane perpendicular to the surface along the diagonal axis CC'. Since the width of the transducer relative to this axis is the measure of sensitivity along the axis, we see that we no longer have uniform

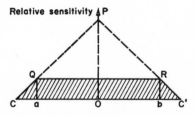

Fig. 6.9. Effective sensitivity distribution along the diagonal axis of a rectangular transducer.

sensitivity; the distribution of sensitivity along the axis CC' is as shown in Fig. 6.9. Although it is uniform over the section ab, it corresponds to a linear taper over the sections Ca and bC'. To work out the directional pattern corresponding to this distribution, it is convenient to regard the latter as made up of a larger triangle CPC' minus a smaller triangle QPR. We then have effectively two linear taper arrays as discussed in Example 5 of the previous section, the response of the smaller being subtracted from that of the larger. We note also that the relative weightings of the two are in the ratio $(CC'/ab)^2$.

Write l_c for the length CC' and l_f for the length ab. Then, using eqn. (6.17) and not bothering about normalization to unity peak height just at present, we have

$$D_c(K) \propto l_c^2 \left[\frac{\sin (Kl_c/4)}{Kl_c/4}\right]^2 - l_f^2 \left[\frac{\sin (Kl_f/4)}{Kl_f/4}\right]^2 \qquad (6.19)$$

$$= \frac{8}{K^2} \left[\cos (Kl_f/2) - \cos (Kl_c/2)\right]$$

$$= \frac{16}{K^2} \sin \frac{K}{4} (l_c - l_f) . \sin \frac{K}{4} (l_c + l_f) \qquad (6.20)$$

We now note that

$$l_c = l_1 \cos \phi + l_2 \sin \phi$$

and

$$l_f = l_1 \cos \phi - l_2 \sin \phi$$

where ϕ is the angle between the diagonal CC' and AA'; thus

$$l_c + l_f = 2l_1 \cos \phi \text{ and } l_c - l_f = 2l_2 \sin \phi$$

Equation (6.20) therefore becomes

$$\frac{16}{K^2} \sin [(l_1 K \cos \phi)/2] . \sin [(l_2 K \sin \phi)/2] \qquad (6.21)$$

At this stage we observe that the peak height of the pattern must be the same (since it occurs for the direction normal to the surface) for all axes, and we have taken it as unity in eqns. (6.18a) and (6.18b). We must therefore normalize this pattern to unity peak height. But in eqn. (6.19) we have a peak height (when $K = 0$) of $l_c^2 - l_f^2$. Now

$$l_c^2 - l_f^2 = (l_c - l_f) (l_c + l_f) = 4l^1 l^2 \cos \phi \sin \phi.$$

Therefore, dividing eqn. (6.21) by this amount, we have

$$D_c(K) = \frac{\sin [(l_1 K \cos \phi)/2]}{(l_1 K \cos \phi)/2} . \frac{\sin [(l_2 K \sin \phi)/2]}{(l_2 K \sin \phi)/2} \qquad (6.22)$$

as the normalized directional pattern in the plane through the diagonal.

If the rectangular transducer were in fact square, so that $l_1 = l_2 = l$, then the sensitivity pattern on the diagonal would be merely a simple linear taper, $\cos \phi = \sin \phi = 1/\sqrt{2}$, $l_c = \sqrt{2}.l$, and $l_f = 0$, so that the directional pattern on the diagonal axis would be

$$D_c(K) = \left[\frac{\sin (l_c K/4)}{l_c K/4} \right]^2 \qquad (6.23)$$

This, of course, agrees with the result for the linear taper line transducer in the previous section—eqn. (6.17)—when normalized for unity peak height.

The method used above is not confined to the diagonal axis, but is quite general. We can, of course, think of θ (which is contained in K) and ϕ as elevation and azimuth angles relative to the transducer surface and its long axis AA'. But if instead we think of $K \cos \phi$ and $K \sin \phi$ as angular coordinates in the orthogonal planes through the axes AA' and BB' respectively, then we see that the directional pattern on the diagonal axis is the product of two directional patterns, one representing the response in terms of $K \cos \phi$ for a line transducer coinciding with AA', and the other the response in terms of $K \sin \phi$ for another line transducer coinciding with BB'. It can be shown that all directional patterns for any axes of a rectangular transducer can be expressed in this way as the product of the patterns of two orthogonal line arrays.

2.2.2. Circular Transducer

Another commonly used planar transducer is the circular disc shown in Fig. 6.10. Again, assume uniform sensitivity over the surface. Owing to the circular symmetry, the directional response is evidently the same for all planes normal to the surface intersecting the disc along a diameter. Take an arbitrary diametrical axis as shown. Then at any point distant r from the centre, along this axis, the relative sensitivity is given by the half-width (h) of the disc along a line normal to the diameter at r.

Now $h^2 = l^2/4 - r^2$, so that using eqn. (6.7),

$$D(K) = \int_{-l/2}^{l/2} (l^2/4 - r^2)^{1/2} \cos Kr \cdot dr \qquad (6.24)$$

and normalizing for unity peak height this can be shown to be

$$D(K) = 2\frac{J_1(Kl/2)}{Kl/2} \qquad (6.25)$$

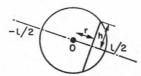

Fig. 6.10. Circular transducer.

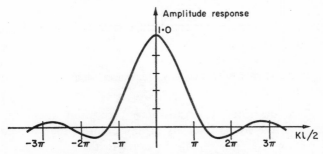

Fig. 6.11. Directional pattern for circular transducer.

where $J_1(Kl/2)$ is the Bessel function of the first kind and first order and of argument $Kl/2$. A graph of this directional pattern is shown in Fig. 6.11.

2.2.3. *Other Planar Transducers and Arrays*

It should now be clear how these can be dealt with, and it is not thought necessary to work out any more examples here.

2.3. *Reciprocity*

In Chapter 5 the equivalent circuits of transducers were discussed, and it is evident that in the ordinary circuit conception of the Reciprocity Theorem,[2] reciprocity applies to transducers. But it applies to transducers and arrays in a wider sense also, since in all cases of transducers and arrays mentioned in this book, the directional patterns determined for a received signal apply also on transmission under the same conditions. Thus, if an array is used to transmit a signal $\cos \omega_q t$, all parts of the array having an excitation (and phase) defined by the same $T(r)$ function

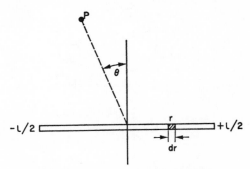

FIG. 6.12. Line array for transmission.

as used to describe its sensitivity on reception, then the dependence of the sound intensity at any point in the far field on the direction of a line joining the point to the array is identical with the directional pattern on reception. For example, consider the line array used for the initial work in Section 2.1 and shown for reception in Fig. 6.1. It is shown as a transmitter in Fig. 6.12. It is now assumed that the function $T(r)$ describes the excitation of the array at a point distant r from the centre. Consider a point P which is so far from the array that it subtends the same angle θ to the normal at any point on the array. (Obviously the diagram cannot correctly show this distance.) Then the contribution to the sound pressure at P due to the elementary length

dr of the array at distance r from the centre is proportional to
$T(r)$. dr . cos $(\omega_q t - Kr)$ where K is defined as before, and the
phase of the signal contribution from the central element of the
array is taken as reference zero. The total sound pressure at P is
proportional to

$$\int_{-\infty}^{\infty} T(r) . \cos (\omega_q t - Kr) . \, dr$$

which is exactly the same as eqn. (6.2) for reception, and the
directional dependence of the magnitude of this is clearly expressed by exactly the same Fourier transform relationships as
given in eqns. (6.4), (6.6), (6.7) and (6.9) for reception.

3. Deflection of Beams

3.1. *General Considerations*

Normally acoustic transducers and arrays are designed to have a
directional pattern with its main beam at right-angles to the
surface or line of the array, and they are then called *broadside*
arrays. It is, in principle, possible to design a line array to have its
main beam in the direction of the line of the array; it is then
called an *endfire* array. But although endfire arrays have important
uses in radio, they are not very practicable in acoustics, and so
we shall continue to confine our attention to arrays which are
basically of the broadside type.

Although an array is basically of the broadside type, nevertheless it is often desired to deflect its beam to point in some other
direction without having to move the array physically. This can be
done by adjusting the phase relationships of the elements of the
array. Since it is almost always necessary to make this adjustment
by means of electrical networks, it is clear that separate electrical
terminals have to be provided for each element of the array, and
thus a continuous transducer is unsuitable for this treatment unless
it can be divided into separate sections. Electrical beam deflection can be applied to arrays of any shape and for deflection in

any direction; but, for simplicity, we shall here confine our discussion to line arrays. The deflection concerned is therefore in a plane containing the array.

In the symmetrical arrays discussed in Section 2.1. the beam was normal to the line of the array simply because it was for this direction that the signals to or from each element of the array added exactly in phase. Clearly, therefore, the beam can be made to point at an angle γ to the normal, if phase-shift networks are so connected to the various elements that all signals add in phase for this angle instead of for the normal.

3.2. Linear Point Array

Consider two adjacent elements of a point array, namely, those numbered s and $s + 1$, with a relative phase-shift of ϕ inserted between them. Assuming reception of a signal at angle θ to the normal, and that the signal wave is $\cos \omega_q t$ at the centre of the array, the signals from these two elements are

$$\cos \left(\omega_q t + \frac{2s\pi d}{\lambda} \sin \theta \right)$$

and

$$\cos \left[\omega_q t + \frac{2(s + 1) \pi d}{\lambda} \sin \theta + \phi \right]$$

The phase difference between adjacent elements is thus

$$\left[\frac{2\pi d}{\lambda} \sin \theta \right] + \phi$$

$$= Kd + \phi \text{ in compact form,}$$

$$= \frac{2\pi d}{\lambda} \left(\sin \theta + \frac{\phi\lambda}{2\pi d} \right) \qquad (6.26)$$

Therefore the directional pattern becomes, corresponding to eqn. (6.13),

$$D(K) = \frac{\sin [n(Kd + \phi)/2]}{\sin [(Kd + \phi)/2]} \qquad (6.27)$$

and it is clear from eqn. (6.26) that if the system is a narrow-beam one (i.e. the array is long in terms of wavelengths) and that we are concerned only with the central part of the pattern, i.e. with the region in which θ is small and $\sin \theta \simeq \theta$, the angle of deflection of the beam is

$$\gamma \simeq \frac{\phi\lambda}{2\pi d} \tag{6.28}$$

Evidently for wider-beam systems, or for very large deflections, the sine relationship means that the deflected beam is wider than the normal beam.

In dealing with the deflection of the beam of a linear point array above, we have assumed that it is the central period of the repetitive directional pattern which is of interest. The pattern as a whole is deflected by the phase shifts and on the K-scale there is no distortion of the pattern as it is deflected. But in terms of the actual angles, the pattern changes its shape on deflection.

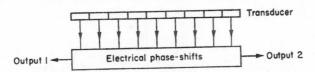

FIG. 6.13. Arrangement of sectionalized transducer for beam deflection.

3.3. Linear Array of Elements each of Finite Length

If each element of the array has finite length (assumed the same for each) then we saw in Section 2.1 that the directional pattern is the product of the array factor and the directional pattern of one element, as indicated by eqn. (6.16). If the elements are connected through phase-shift networks in order to get a deflection of the beam, then clearly it is only the array factor which is shifted along the K-axis as described in the previous section; the pattern of the individual elements obviously remains unaltered. Thus the overall directional pattern changes very considerably as it is deflected.

A most important case is that where $a = d$, i.e. we have a continuous line array merely divided into sections in order that phase-shifts can be inserted for deflecting the beam. This case is illustrated in Fig. 6.13 for a nine-section transducer and for signal reception. Evidently, reversal of the arrows makes the

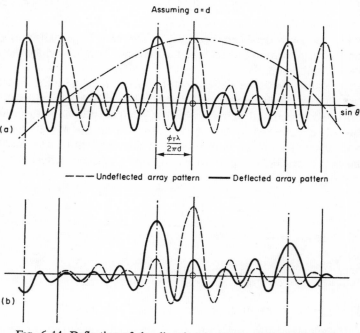

FIG. 6.14. Deflection of the directional pattern of a sectionalized line array showing (a) components of deflected pattern and (b) resultant pattern.

system applicable for transmission. When no phase-shifts are added, the directional pattern is merely the usual [sin $(Kl/2)]/(Kl/2)$, which is given by the product of the pattern of one element and the undeflected array-factor pattern, shown in Fig. 6.14. When phase-shifts are added, the array-factor pattern is shifted as shown in Fig. 6.14(a), and the overall pattern becomes that shown in Fig. 6.14(b). No longer do the zeros of the pattern of one section coincide with the main repetitive peaks of the array-factor pattern, with the result that larger undesirable minor lobes are introduced by the deflection process, and these get larger as the deflection is increased. The limit is reached when the array-factor pattern is shifted one half-period. After that, the patterns merely repeat those previously obtained.

It is apparent that the number of sections into which the transducer is divided is important. A larger number of sections means n is increased and d is decreased. Thus the repetition period of the array-factor pattern is widened, and there are more minor lobes in any one period of the pattern. This means that the beam can be deflected very much further before the limit is reached; and for any given deflection the distortion is much less. The choice of n is therefore determined by the maximum deflection required, and by the permissible distortion of the pattern which may conveniently be specified by the maximum acceptable height of any minor lobe.

In the limit we have $n \to \infty$, $d \to 0$, and for angular deflections which are not too large, we can obtain a simplified expression for the deflected overall directional pattern. Thus, since as $a \, (= d) \to 0$,

$$\frac{\sin (Ka/2)}{Ka/2} \to 1$$

and

$$\sin [(Kd + \phi)/2] \to (Kd + \phi)/2$$

then, using eqns. (6.16) and (6.27),

$$D(K) \to \frac{\sin [(Kl + \phi_T)/2]}{(Kl + \phi_T)/2} \qquad (6.29)$$

where, comparing with eqn. (6.27), a normalizing factor of $1/n$ has been introduced to bring the pattern to unity peak height, and $l = nd$, while $\phi_T = n\phi$ and is therefore the *total* phase-shift in the network in Fig. 6.13. This pattern is merely a deflected $(\sin x)/x$ pattern, deflected by an amount $\phi_T/2$ on the scale of x.

In Section 4 of this chapter we shall find it convenient to work with x in place of $Kl/2$. A graph of the pattern deflected by an amount corresponding to $\phi_T = \pi$ is shown in Fig. 6.15.

The patterns can be deflected to one side or the other of the normal according to which of the two outputs shown in Fig. 6.13 is taken.

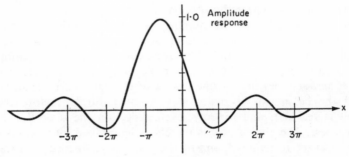

FIG. 6.15. (Sin x)/x pattern deflected by $\pi/2$ so that $\phi_T = \pi$.

4. Synthesis of Directional Patterns

4.1. *General Considerations*

For simplicity we shall once again restrict the discussion to line arrays, although many of the conclusions reached apply, in more complicated form, to two-dimensional arrays.

It was mentioned in Section 2.1 that, in principle, the Fourier transform relationship between the taper function $T(r)$ and the directional function $D(K)$ solves the design problem of arrays, but that, in practice, it helps very little. It is easy to see why this is. The inverse transform

$$T(r) = \frac{1}{2\pi} \int\limits_{-\infty}^{\infty} D(K) \cos Kr \, . \, \mathrm{d}K \qquad (6.30)$$

specifies the taper function required to give a particular desired (symmetrical) directional function. Now, firstly, the integration has to be carried out over all possible values of K, yet $D(K)$ is not likely to be defined by the user of the equipment over ranges of K which give $\sin \theta > 1$, i.e. for directions outside the range of real angles. But the values of $D(K)$ outside the range of real values of θ may be quite significant, as will be obvious for the case when wide beams are considered, since then quite large minor lobes will be outside this range and, of course, when point arrays are used with their repetitive patterns; and if superdirectivity (see Section 4.4) is to be obtained, then even the largest or main lobes may be placed in the range of $\sin \theta > 1$. So inadequate information is normally available to enable the integration to be done.

A second difficulty with eqn. (6.30) is that it will result in values of $T(r)$ being obtained for all values of r. But in practice an infinitely-long array cannot be used; a maximum length will be specified by the user for various reasons. It is clear, then, that eqn. (6.30) does not, in itself, give any basis for design, i.e. it cannot say what is the best taper function to meet a particular directional requirement.

Another approach to the synthesis of a required directional pattern is that due to Schelkunoff.[3] It applies to a linear point array with equally spaced elements. The freedom to obtain an approximation to a desired directional pattern is obtained by allowing each element to have, if necessary, a different sensitivity and phase. For such an array with element spacing of d, and numbering its elements from 1 at one end to n at the other, the response to an applied acoustic wave which is sinusoidal and has a reference phase of zero at one end is the vector sum of the signal amplitudes received on each successive element, namely:

$$D(K) = A_0 + A_1 \exp{(jKd)} + A_2 \exp{(j2Kd)} + A_3 \exp{(j3Kd)}$$
$$+ \ldots + A_{n-1} \exp{[j(n-1)Kd]} \quad (6.31)$$

where A_s is the complex sensitivity of the element numbered $s + 1$. Now this is a polynomial in $\exp{(jKd)}$ and can be expressed as the product of $(n-1)$ factors, i.e.

$$D(K) = [a_1 - b_1 \exp{(jKd)}][a_2 - b_2 \exp{(jKd)}]$$
$$\ldots [a_{n-1} - b_{n-1} \exp{(jKd)}] \quad (6.32)$$

Each of these factors defines the position of a zero. If they correspond to real values of K they lie on the K-axis as we have previously used it in drawing directional patterns; otherwise they lie in a complex K-plane and do not give measurable zeros of response, although their presence may be indicated by dips in the pattern. Clearly, with only n elements in the array, the directional pattern can be specified with only $(n - 1)$ degrees of freedom, and the specification of $(n - 1)$ zeros of the desired pattern defines the pattern exactly. Thus if we put the zeros where we think they will make an optimum pattern, we can then calculate, by putting eqn. (6.32) into the form of eqn. (6.31), the values of the sensitivity and phase required in each element of the array.

This method is very far from straightforward and involves very laborious calculations. A much easier and, in the long run, more direct method of synthesis is given below.

4.2. *Synthesis Using Superposed* (sin x)/x *Patterns*

This method, first proposed by Woodward[4, 5] for radio aerials, is applicable strictly only to continuous transducers, but can be applied to point arrays provided the elements are uniformly spaced and are sufficiently numerous (and therefore close together) to cause the repetitive main lobes of the directional pattern to fall either well away from the angular sector of interest, or, preferably, outside the range of real angles. The difference in the patterns between eqns. (6.10) and (6.13) is then negligible. The method is based on the superposition of a number of (sin x)/x patterns deflected by different amounts to one or other side of the $x = 0$ axis, and with peak amplitudes chosen individually. If individual patterns are spaced at intervals of π on the x-scale, then the peak of each falls at the zeros of all the others. If the amplitude scale of the pattern with peak at $x = s\pi$ is made equal to the required height of the directional pattern at $x = s\pi$, then the required curve will be matched exactly at n points if n component patterns are used. The value of the curve at intermediate points will naturally not be exactly what is required. The whole process is very easily and quickly carried out, with practically no calculation, by a simple graphical process.

The method is illustrated in Fig. 6.16. It can be expressed mathematically thus:

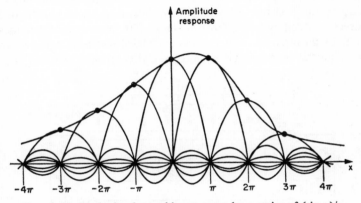

FIG. 6.16. Synthesis of an arbitrary curve by a series of $(\sin x)/x$ curves.

For an even number of component curves, $n' = 2m$,

$$D(x) = \sum_{s=-m}^{s=m-1} A_s \frac{\sin [x - (2s + 1) \pi/2]}{x - (2s + 1) \pi/2} \qquad (6.33)$$

and for an odd number, $n' = 2m + 1$

$$D(x) = \sum_{s=-m}^{s=m} A_s \frac{\sin (x - s\pi)}{x - s\pi} \qquad (6.34)$$

It is assumed here that there is an equal number of matching points on each side of the $x = 0$ axis, but this is not a necessary limitation.

Now the basic $(\sin x)/x$ patterns are readily generated by considering a continuous uniform line transducer divided into n elements and connected to a series of phase-shifting networks (or "delay lines", as they may often be called since they usually comprise series inductance and shunt capacitance), each one as in Fig. 6.13. If it were desired to try this synthesis by a practical method instead of a graphical method, the set-up of Fig. 6.17 could be used. Each delay line has a total phase-shift of 2π rad more than the previous line. If n' is even, then the total phase-shift in the first line is π rad, but if n' is odd, there will be one directly added output with no phase-shift, and the total phase-shift in the first line will be 2π rad. The directional pattern can be recorded or displayed by mechanical rotation of the array at

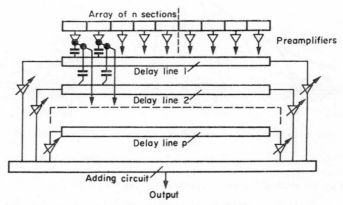

Fig. 6.17. Delay-line system for the synthesis of directional patterns.

a speed synchronized to that of the display, with a signal from a distant point source being received at the array. The amplitude adjustments A_1, A_{-1}, etc., may be adjusted empirically or systematically to give desired values of the directional pattern at the appropriate points. Correction must be made for the successive reversals of phase of the signal at the outputs of the successive delay lines, since it is the patterns which are to be added, and the carrier phase must be the same for each. But it will naturally be necessary to reverse the phase when one of the component patterns is to be subtracted, rather than added.

It has already been pointed out in Section 3.3 that the limit of deflection of a pattern is when the array-factor pattern is deflected by one half-period. This means that if there are n sections in the transducer, then there are n possible patterns which can be available for superposition. Clearly, therefore, if n' patterns are to be used in the synthesis, it is necessary to divide the transducer into at least n' sections. The optimum arrangement is when $n = n'$. In the physical arrangement of Fig. 6.17, where two patterns are obtained from each delay line by using both ends, the number of delay lines which can be used is limited to $n/2$ if n is even and $(n - 1)/2$ if n is odd. There is, however, the additional point that, as stated in Section 3.3, the deflected curves approximate to deflected $(\sin x)/x$ patterns only if n is large. Thus if n is too small, the result is not accurate.

The use of actual phase-shift networks or delay lines in a transducer system is unavoidable if an unsymmetrical directional pattern is needed, although they need not be arranged as in Fig. 6.17. The use of an amplitude control and phase network in the circuit from each array element before their outputs are added can give exactly the same result, although a good deal of calculation is needed to obtain the necessary values of the complex taper function.

If, as is more usual, a symmetrical pattern is required, there is no need to use the delay lines or any other phase-shift networks in the physical realization of the system. They serve merely as one stage of the "paper exercise" of synthesis. All that is needed in the physical system is to adjust the amplitude response of the individual elements of the array before adding their outputs together. The way the necessary taper function is calculated from the pattern synthesis is illustrated in the examples which follow.

EXAMPLE 1. *Directional pattern with reduced minor lobes.* This is a very elementary example, just to bring out some interesting results. It is hardly synthesis in the formal sense.

The type of directional pattern which would probably be considered highly desirable in most practical applications is shown by the full-line curve in Fig. 6.18. This pattern has no minor lobes. A simple approach to such a pattern can be made on the lines described above. Matching of amplitudes is done at $x = \pm \pi/2, \pm 3\pi/2$, etc., and the only matching points at which the response is required to be other than zero are $x = + \pi/2$ and $x = - \pi/2$. Thus only two $(\sin x)/x$ patterns are required for synthesis, and these are identical except for having opposite displacements along the x-axis. This means that in the delay-line

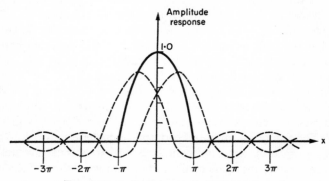

FIG. 6.18. A simple example of synthesis.

system, only one delay line is used, with outputs taken from both ends. For simplicity, at the moment, assume that the amplitude required at $x = \pm \pi/2$ is unity. Then the synthesized directional pattern is

$$D(x) = \frac{\sin (x + \pi/2)}{x + \pi/2} + \frac{\sin (x - \pi/2)}{x - \pi/2} \qquad (6.35)$$

This can immediately be recognized as the same as that obtained by the cosine taper array given as Example 2 in Section 2.1, and is graphed in Fig. 6.4, whence it can be seen that it does partially meet the specification. That it can be obtained by the cosine taper can be proved from the present method of approach quite simply.

Consider the section of the array numbered s, the numbering being outwards from the centre section, which is numbered zero.

Let the signal at its terminals be

$$V \cos (\omega_q t + sKd)$$

At one end of the delay line this particular component of the output becomes proportional to

$$V \cos [\omega_q t + sKd + (m - s) \phi] \qquad (6.36)$$

where m is the number of the outermost section and ϕ is the phase-shift per section of the delay line. At the other end the corresponding output is proportional to

$$V \cos [\omega_q t + sKd + (m + s) \phi] \qquad (6.37)$$

When these two are added together, they become proportional to

$$V \cos s\phi \, . \, \cos (\omega_q t + sKd + m\phi) \qquad (6.38)$$

The phase-shift $m\phi = \pi/2$ rad is a constant phase-shift applied to the signals from all elements (since it is independent of s) and so merely shifts the carrier phase of the final output. The term $\cos s\phi$, however, is the amplitude taper, and since $s\phi = (s/m) \, \pi/2$, this gives exactly the same cosine taper as was used in the earlier example.

It can be seen that, for any symmetrical system, the above process of calculating the taper function always applies, although when more than one "notional" delay line is used in the synthesis, more than one pair of terms like (6.36) and (6.37) have to be added together.

EXAMPLE 2. *Flat-topped beamshape.* The directional patterns so far illustrated have had rounded main peaks. It is obvious that in some practical applications, e.g. in the transmitter of an electronic sector-scanning sonar (see Chapter 1), it would be advantageous to have a flat-topped beam so that constant response is available over the whole sector.

A flat-topped beam can be approximated by the process of synthesis described above. Three equally spaced basic patterns can be added as shown in Fig. 6.19, with one peak at the centre of the desired beam and each of the other peaks at one edge of

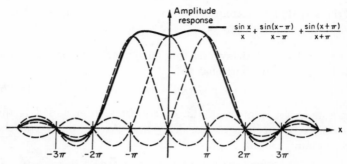

Fig. 6.19. Widening the main lobe (flat-topped beamshape).

it. The length of the array is made such that the basic patterns are spaced apart at intervals of π on the x-scale. Then the sum of the three patterns gives a beamshape which is flat over the specified range to within $\pm 3\%$. The overall directional pattern is evidently

$$D(x) = \frac{\sin(x + \pi)}{x + \pi} + \frac{\sin x}{x} + \frac{\sin(x - \pi)}{x - \pi} \qquad (6.39)$$

and it is easily shown that the taper function required (in place of a physical delay line) is

$$T(r) = 1 - 2 \cos(s/m)\,\pi \qquad (6.40)$$

which is shown as a graph in Fig. 6.20. It will be observed that

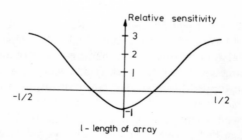

l - length of array

Fig. 6.20. Distribution of sensitivity along the array for flat-topped beamshape.

response of opposite polarity has been introduced at the ends of the array.

4.3. *"Split-beam" Array Systems*

It is often desired to determine the direction of a received signal by means of a null indication rather than by using the peak of a directional pattern. This is not, in general, any more accurate but is sometimes more convenient. A common and simple way of doing this is to divide a continuous uniform array into two halves and subtract the output of one half from that of the other. The output of one half (retaining the centre of the whole array as the phase reference) is proportional to

$$\frac{\sin (Kl/4)}{Kl/4} \cos (\omega_q t + Kl/4)$$

while that of the other is proportional to

$$\frac{\sin (Kl/4)}{Kl/4} \cos (\omega_q t - Kl/4)$$

The difference is therefore proportional to

$$\frac{\sin^2 (Kl/4)}{Kl/4} \sin \omega_q t$$

so that the directional function is

$$D(K) = \frac{\sin^2 (Kl/4)}{Kl/4} \tag{6.41}$$

This can be seen to be the symmetrical directional function of one half-array multiplied by the odd function $\sin (Kl/4)$, and thus it has a null at $K = 0$ with a positive peak to one side and a negative peak to the other, as shown in Fig. 6.21, curve (a). If polarity is ignored (e.g. by use of rectifiers) the pattern consists of the double-peaked curve (b), and it is easily seen how this has given the name "split-beam" to the system.

A split-beam system can also easily be designed by the use of

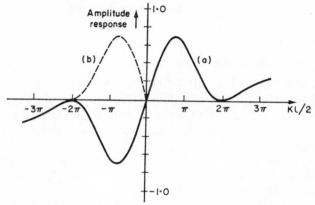

FIG. 6.21. Directional pattern of "split-beam" system, (a) according to eqn. (6.41), (b) ignoring polarity.

the delay-line method described in Section 4.2. If the single delay line used in Example 1 of Section 4.2 has its two outputs subtracted instead of added, then the directional pattern is

$$D(K) = \frac{\sin\left[(Kl-\pi)/2\right]}{(Kl-\pi)/2} - \frac{\sin\left[(Kl+\pi)/2\right]}{(Kl+\pi)/2} \qquad (6.42)$$

which is shown as a graph in Fig. 6.22. It is very similar to the previous pattern. It is easily shown that the required taper function is a "sine taper", $T(r) = \sin\left[(s/m)\,\pi/2\right]$; this means that the two halves of the array are connected in opposite polarity, but do not have uniform sensitivity over their length as in the first split-beam case.

4.4. Superdirective Arrays

In the synthesis of directional patterns, a greater degree of control of the resultant pattern may be obtained by including among the constituent patterns some in which the main peak occurs at x-values which lie outside the range of real angles. Then, in the practical system, only the minor responses of these patterns

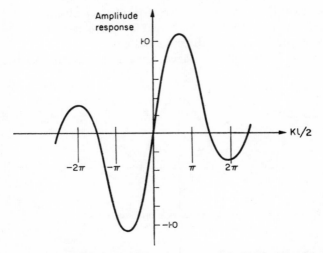

Fɪɢ. 6.22. "Split-beam" directional pattern obtained using delay
line.

contribute to the resultant pattern and their main lobes will not
be observable. Such arrays can be made to have narrower beamwidths *and* smaller minor lobes than the ordinary arrays, and are
consequently known as superdirective arrays. But the deflection
of constituent patterns outside the range of real angles can
readily be seen to necessitate a spacing of elements (or transducer
sections) at less than half-a-wavelength, and the taper functions
involved are always oscillatory, involving the excitation (or
response) of some adjacent elements in opposite phase. Moreover, the excitation (or response) of individual elements has to
be very much greater (often enormously greater) than the resultant, so that the system is inefficient and/or noisy. For these
reasons, superdirective arrays are very difficult to realize in
practice, and very unattractive in spite of their good theoretical
directivity. No application in an operational system is known to
the authors.

Although superdirective arrays are thus not of great practical importance, they are of very great academic interest, and their literature is huge.[6] We shall therefore give here briefly three simple examples of superdirective designs.

In Fig. 6.23, the full-line curve shows a normal $(\sin x)/x$ pattern

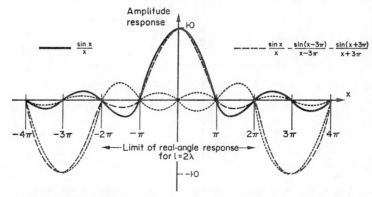

FIG. 6.23. Synthesis of a superdirective response with reduced minor lobes and slightly-narrowed main beam.

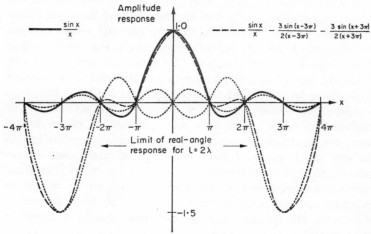

FIG. 6.24. As for Fig. 6.23 but with minor lobes practically eliminated.

of unity peak height. It will be assumed that the length of the array is twice the wavelength, so that the limit of real-angle response is given by $x = \pm 2\pi$. Within this range fall the first minor lobes of the pattern with an amplitude of 0·217. These minor lobes can be reduced to a very small value, with an accompanying narrowing of the main lobe, by adding in opposite sense the two deflected patterns shown by dotted lines. Each of these has a peak height of $-1·0$ occurring at $x = \pm 3\pi$, i.e. outside the range of real angles. It is seen that the resultant pattern shown in dashed lines is, in the range of real angles, very considerably

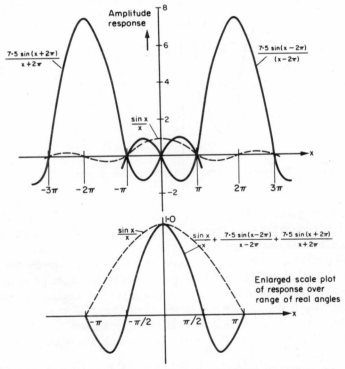

Fig. 6.25. Example of synthesis of superdirective response with main beam greatly narrowed.

improved. Similarly, in Fig. 6.24, two patterns of peak amplitude −1·5 have been added to the ordinary pattern with still further improvement.

Fig. 6.25 shows an example of a slightly different kind. Here it is assumed that the array has a length of one wavelength, so that the limit of real-angle response is given by $x = \pm \pi$. The aim has been to obtain a beam of one-third the width, in the θ-scale, of the ordinary beam corresponding to this length of array, or on the x-scale with a width of one-half that of the

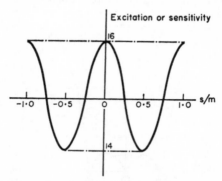

FIG. 6.26. Taper function of superdirective response shown in Fig. 6.25.

ordinary pattern. This has been accomplished by adding two deflected patterns with their peaks at $x = \pm 2\pi$, and amplitudes calculated to give a zero resultant at $x = \pm \pi/2$. This necessitates the amplitude of the peaks outside the range of real angles being 7·5 times that of the ordinary pattern. The resultant pattern is seen to have minor lobes, in the range of real angles, of peak amplitude 0·44, i.e. rather larger than those of the ordinary pattern; but the main beam is narrowed as desired.

In this last example, it is easily shown that the taper function required is

$$T(r) = 1 + 15 \cos (2s/m) \pi \qquad (6.43)$$

which is graphed in Fig. 6.26. It is clearly oscillatory, with excitations as high as 16 times the resultant response. The minimum number of sections in the transducer with which it can be even approximately realized is evidently five.

4.5. Directivity Index

For many purposes it is most suitable to define the directivity of an array in terms of the beamwidth (normally measured at the half-power points, i.e. the points at which the response is 3 dB below the peak) and of the maximum relative height of minor lobes. But a single figure which applies irrespective of the shape of the pattern and is equally applicable to beams produced by line or two-dimensional arrays is the *directivity factor*. In terms of a transmitter it is defined as the ratio of the maximum transmitted intensity to the intensity which would result from the same transmitted power distributed uniformly in all directions. In terms of a receiver it is the ratio of the output power developed by a signal in the direction of maximum response to that developed by the same signal if it were uniformly distributed over all directions. Referring to the pattern of a line array for simplicity, the directivity factor (D.F.) may be expressed thus:

$$\text{D.F.} = \frac{[D(0)]^2}{\frac{1}{2\sin\theta}\int\limits_{-\pi/2}^{+\pi/2}[D(\theta)]^2 \, \sin\theta \, \mathrm{d}\theta} \qquad (6.44)$$

where $D(\theta)$ is the directional pattern expressed as a function of the physical angle θ and not of K or x. $D(0)$ is the value of $D(\theta)$ for the peak response at $\theta = 0$. The *directivity index* (D.I.) is this ratio expressed in decibels, i.e. $\text{D.I.} = 10 \log_{10} \text{D.F.}$

4.6. Dependence of Directional Patterns on Frequency

It will be appreciated that in all the previous work in this chapter the signal has been assumed to be a single sinusoid at frequency ω_q. In practice, however, as explained in Chapter 2, the bandwidth of the system has to be sufficient to pass a signal which may be of pulse form, or of frequency-modulated form, or of a random form (as in the observation of fish noise, for example). The question thus arises as to how dependent the directional patterns are on frequency.

With the simplest arrangements, for example, the continuous uniform transducer and the uniform point array, the directional pattern clearly does not change in *shape* with frequency, if expressed in terms of sin θ and not θ, but since $K = 2\pi/\lambda$ sin θ, the *scale* of the pattern is inverse with respect to wavelength. Thus as the frequency increases the whole pattern closes up and the beam narrows in inverse proportion to frequency, while additional minor lobes fall into the range of real angles. Since in a typical sonar system the frequency bandwidth is usually not more than one-fifth to one-tenth of the centre frequency, this amount of dependence of directivity on frequency is usually of no significance.

The conclusion reached above can be seen to apply also to all other arrangements which:

(a) Do not use physical phase-shift networks, the responses of which are, of course, dependent on frequency, and

(b) Do not have large responses in the range of K or x corresponding to sin $\theta > 1$, since these responses will enter the range of real angles as the frequency increases.

These conditions mean, in effect, that the conclusion applies to all symmetrical and non-superdirective arrays.

In the case of superdirective arrays, the range of frequency over which they can operate is limited not only by the need to keep relatively enormous responses right outside the range of real angles, but also by the practical difficulty that the relative amplitude and phase response of the various elements cannot be kept sufficiently constant over a range of frequencies. Since the overall response is dependent on small differences between large individual-element responses, precision in the latter is obviously essential although in practice not readily obtainable.

5. Signal-to-noise Ratio in Transducer Arrays

This is a complicated subject on which much has been written.[7, 8, 9, 10] In this book we can give only a very brief outline of some of the more important and fundamental ideas. The ideas of noise and its mathematical description were outlined in Chapter 2, and it was made clear that the chief characteristic of noise is its randomness. As a result of this, when more than one source or receiver of noise is concerned, the question of correlation of the noise waveforms becomes important. With this in mind we can consider, firstly, noise arising in the water, and, secondly, noise arising in the receiving transducers and circuits themselves. When we refer to the signal, we shall always assume a signal arriving from a single distant point source in the direction of the axis of the beam of the directional pattern.

5.1. Noise Arising in the Water

As explained in Chapter 1, noise arises in the water due to numerous man-made and natural causes. If the noise sources are numerous, uncorrelated with one another, and give the same r.m.s. sound pressure at the receiver irrespective of their direction, then it is clear that the ability of the receiver to discriminate between a signal coming from a single direction and the noise is measured by its directivity factor or index, eqn. (6.44), insofar as this is constant over the frequency bandwidth of the noise or that of the array (whichever is the smaller).

5.2. Noise Arising in the Transducers

Noise arises in the transducers due to the thermal-agitation noise of their loss resistance. The thermal-agitation noise of the water itself can be considered as the noise of the radiation resistance and included in this section rather than the previous one, but it must be realized that for this purpose the temperature of the radiation resistance has to be taken as that of the water, while that of the loss resistance is that of the transducers.

It is clear that the noise of the loss resistance is additional noise added to the system and thus our first conclusion is that the signal-to-noise ratio of the receiver is dependent on the efficiency of the transducers, although as the total water noise usually exceeds the noise due to the loss resistance, this dependence is often not very marked.

In addition to this factor, however, the signal-to-noise ratio is also affected by the size and by the taper function of the transducer or array. For example, in a symmetrical array the signal voltages on the various sections or elements are in phase (since we assume the signal direction is along the axis of the beam) and so add together in the resultant output; but since the noise voltages have random and uncorrelated waveforms, they add only on a power basis, as explained in Chapter 2. Thus if there are ns ($= 2m + 1$) elements or sections, each with a relative response T_s, where s identifies any particular element by numbering from zero at the centre, the total signal voltage is

$$\sum_{s=-m}^{s=+m} T_s \qquad (6.45)$$

while the total noise voltage has a r.m.s. value of

$$\sqrt{\left(\sum_{s=-m}^{s=+m} T_s^2 \right)} \qquad (6.46)$$

It is clear that the signal-to-noise ratio is improved 3 dB for every doubling of the number of elements, or, in continuous transducers, for every doubling of the area of the active face.

The relative signal-to-noise performance of different taper functions can be compared by using a *noise figure* (N.F.), usually expressed in decibels, defined thus:

$$\text{N.F.} = 10 \log_{10} n \cdot \frac{\displaystyle\sum_{s=-m}^{s=+m} T_s^2}{\left(\displaystyle\sum_{s=-m}^{s=+m} T_s \right)^2} \text{ dB} \qquad (6.47)$$

The factor n is introduced to make the N.F. a comparison only of the effect of different taper functions and not of the length of the array; otherwise it would obviously be better for longer arrays than for shorter ones.

For continuous transducer arrays, the N.F. can be expressed in integral form, thus:

$$\text{N.F.} = 10 \log_{10} l \cdot \frac{\int_{-\infty}^{\infty} [T(r)]^2 \, dr}{\left(\int_{-\infty}^{\infty} T(r) \cdot dr \right)^2} \, dB \qquad (6.48)$$

It is clear that this criterion of signal-to-noise performance effectively takes the uniform array as reference, since for this array, N.F. = 0 dB. A worse noise performance is indicated by a larger N.F. For some other arrays considered in this chapter, the noise figures are as shown in the table below. Easy ways of calculating noise figures are described in reference 7.

Name of arrangement	Section of chapter in which discussed	Taper function, $T(r)$, over range of r from $-l/2$ to $+l/2$	N.F. (dB)
Cosine taper	2.1	$\cos (\pi r/l)$	0·92
Linear taper	2.1	$1 - 2 \mid r \mid /l$	1·26
Circular taper	2.2.2	$1 - 4r^2/l^2$	0·34
Flat-topped beam	4.2	$1 + 2 \cos (2\pi r/l)$	4·28

A special comment is needed on superdirective arrays. Since the overall signal response is comprised of differences between large individual element responses yet the r.m.s. noise voltages all add together, we obtain very poor, i.e. very high, noise figures. If we take the example given in Section 4.4, where the taper function was given in eqn. (6.43) and the directional pattern in Fig. 6.25, we find that N.F. = 20·5 dB, which is a very poor noise performance indeed. Yet the directivity index was only approximately 3 dB better than that of the uniform array of the same length. This is typical; a very high price is paid in superdirective arrays for the improvement in directivity.

It is clear that no array can give a better signal-to-noise performance than a uniform array.

Noise arising in the circuits associated with each individual element may obviously be included with the thermal noise of the loss resistance for the above comparative calculations.

5.3. *Relation between Directivity Index and Noise Figure*

Provided we are concerned with long arrays, i.e. $l/\lambda \gg 1$, so that the individual elements or section-centres are not very close together (say $d/\lambda > 1$), then it may be shown that the noise arising in the medium, if distributed spatially as specified in Section 5.1, gives waveforms in the various elements which are largely uncorrelated one with another. In these circumstances the N.F. is evidently some sort of measure of the directional discrimination of the array against noise in the water, and it is clear that the directivity index (D.I.) and the N.F. must be related.

Now since the basic relationship between the $D(K)$ and $T(r)$ functions is the Fourier transform discussed in Section 2.1, it follows from the Fourier integral energy theorem, often called Rayleigh's Theorem,[1] that

$$\int_{-\infty}^{\infty} |T(r)|^2 \, dr = \frac{1}{2\pi} \int_{-\infty}^{\infty} |D(K)|^2 \, dK. \tag{6.49}$$

This is really a statement of the fact that the power in the array equals the power in the signal. Evidently, for this purpose, both $T(r)$ and $D(K)$ must be expressed in terms of the acoustic signal or both in terms of the electrical signal at the terminals; i.e. the transducer losses must not be included in one and omitted from the other.

From eqn. (6.7) or eqn. (6.9), assuming a symmetrical array,

$$\int_{-\infty}^{\infty} T(r) \cdot dr = D(0) \tag{6.50}$$

Therefore, from eqn. (6.48),

$$\text{N.F.} = 10 \log_{10} \frac{l}{2\pi} \cdot \frac{\displaystyle\int_{-\infty}^{\infty} [D(K)]^2 \, dK}{[D(0)]^2} \qquad (6.51)$$

From this it can be seen that if the effective directional response of the array is confined to relatively small values of θ, i.e. it is a long array, so that over the effective range of θ we can put $\sin \theta \simeq \theta$, then $\theta \propto K$, and the main term of the N.F. is merely the inverse of the main term of the D.I. as given by eqn. (6.44). Thus the difference (in decibels) between the D.I. of one array and that of another of the same length is exactly the same as that between their N.F.s. Apart from purely numerical factors involving the length of the arrays the two criteria measure the same thing. Although this is fundamentally interesting, we must remember that it applies only to long arrays, and does not apply to super-directive arrays, where the elements are necessarily spaced less than a half-wavelength apart. It is worth pointing out there are also other kinds of array, not discussed in this book (e.g. multiplicative arrays[11]) in which this relationship fails. It should, moreover, be remembered that the D.I. is frequency-dependent, so that the relationship is really confined to narrow-bandwidth systems.

References

1. For example, STUART, R. D., *An Introduction to Fourier Analysis*, Methuen, London (1961).
2. See any textbook on electrical network theory, e.g. TUCKER, D. G., *Elementary Electrical Network Theory*, Pergamon, Oxford (1964).
3. SCHELKUNOFF, S. A., A Mathematical Theory of Linear Arrays, *Bell System Tech. J.* **22**, 80 (1943).
4. WOODWARD, P. M., A Method of Calculating the Field over a Plane Aperture Required to Produce a Given Polar Diagram, *J. Inst. Elect. Engrs.* **93**, Part IIIA, 1554 (1946).

5. WOODWARD, P. M., and LAWSON, J., The Theoretical Precision with which an Arbitrary Radiation Pattern may be obtained from a Source of Finite Size, *J. Inst. Elect. Engrs.* **95**, Part III, 363 (1948).
6. For a comprehensive bibliography see BLOCH, A., MEDHURST, R. G., and POOL, S. D., Super-directivity, *Proc. Inst. Radio Engrs.* **48**, 1164 (1960).
7. TUCKER, D. G., The Signal/Noise Performance of Electroacoustic Strip Arrays, *Acustica* **8**, 53 (1958).
8. TUCKER, D. G., Signal/Noise Performance of Super-Directive Arrays, *Acustica* **8**, 112 (1958).
9. WELSBY, V. G., The Signal/Noise Gain of Ideal Receiving Arrays, *Inst. Elect. Engrs.* Monograph 470E (Sept. 1961).
10. TUCKER, D. G., Signal/Noise Performance of Multiplier (or Correlation) and Addition (or Integrating) Types of Detector, *Inst. Elect. Engrs.* Monograph 120R (Feb. 1955).
11. WELSBY, V. G., and TUCKER, D. G., Multiplicative Receiving Arrays, *J. Brit. Inst. Radio Engrs.* **19**, 369 (1959).

Problems and Solutions

THE problems set out below are of varied nature and complexity; some are typical examination questions but many are designed mainly to extend the student's knowledge of the subject. It has therefore been considered desirable to give fairly full solutions. The problems are grouped according to the chapters on which they are based, as indicated by the numbering system.

Problems

2.1. Calculate the resonant frequency and 3-dB bandwidth of the circuit shown in Fig. P1 assuming the overall effective Q-factor considerably exceeds unity. Calculate also the response–time curve of this circuit when a tone of the resonant frequency is suddenly applied, and see how the rate of rise corresponds to the bandwidth according to Section 2 of Chapter 2.

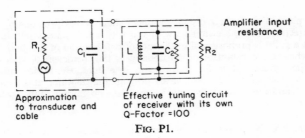

FIG. P1.

2.2. Two closely spaced hydrophones receive the same signal from a distant source. The spacing is such that while the signal waves received by the hydrophones are exactly correlated, the water noise at the two hydrophones has a correlation factor of $0 \cdot 5$. The loss resistance and the receiving amplifiers following the hydrophones contribute noise which in each circuit is of the same r.m.s. magnitude as the water noise at the same point in the circuit but which is completely uncorrelated between the two circuits. Calculate how much the signal–noise ratio, when the two hydrophone channels are combined together, differs from that in the water at the hydrophones.

2.3. Use is made frequently of the relationship: "the mean of the sum of a number of random quantities is equal to the sum of the means of each taken separately".
Prove this.

2.4. A number (n) of noise sources, all of equal r.m.s. voltage, and with very narrow bandwidths (all identical), and all uncorrelated with one another, are connected to a delay line as shown in Fig. P2. Assuming that the phase-shift

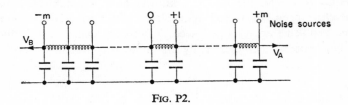

FIG. P2.

ϕ per section of the delay line is constant over the narrow band of the noise, determine the values of ϕ which make the resultant voltage waveforms (V_A and V_B), at the two ends of the delay line, completely uncorrelated with one another.

3.1. A plane Y-cut quartz crystal produces shear waves propagating along the x-direction. Using Newton's law of viscosity which states that the shear stress at a point is proportional to the rate of deformation of the fluid at that point, where the constant of proportionality is the coefficient of viscosity μ, derive the wave equation for shear waves in a fluid. Hence determine the absorption coefficient for shear waves propagating in water at a frequency of 2 Mc/s, and comment on the significance of the magnitude of the coefficient obtained.

3.2. Calculate the phase difference between the acoustic pressure and the particle velocity due to a point source radiating into sea-water at a frequency of 2 kc/s at distances of 10 cm, 1 m and 10 m. What are the corresponding specific acoustic impedances at these distances?

3.3. A sound source consisting of a sphere of radius 1 m radiates spherical waves into sea-water at a frequency of 25 c/s. Derive an expression for the specific acoustic admittance presented by the medium and sketch its equivalent circuit.

If the source is designed to radiate a peak power of 1 kW, determine

(a) the peak values of acoustic pressure, particle displacement, velocity and acceleration at a distance of 10 m from the centre of the source, and

(b) the displacement, velocity and acceleration of the surface of the source.

3.4. Calculate the impedance presented by an acoustic transformer consisting of a $\lambda/4$ section of steel radiating into water. What is the effect of increasing the thickness of the steel section to $\lambda/2$?

3.5. An acoustic doublet consists of two point sources of equal strength M, but in opposite phase, separated by a distance d. Show that the magnitude of the acoustic pressure at a distance r from the doublet, where $r \gg d$, is given by

$$\frac{\rho_0 c k^2 M d \cos \theta}{4\pi r}$$

where θ is the direction of the point of observation relative to the line joining the two sources. Sketch the resulting radiated pressure pattern.

4.1. A 20 kc/s omnidirectional sound generator is immersed in water at a depth of 10 m. If the surface of the water is perfectly flat, the signal received by a second transducer at a depth of 20 m is found to pass through maxima and minima as the separation of the two transducers is increased. Explain this and determine the separation at which the last minimum occurs.

4.2. Two layers of water are in contact at a depth of 60 m, the upper and lower layers having temperatures of 15 and 5°C respectively. If the sound is incident from the upper layer upon the boundary at an angle of 45°, calculate

(a) the angle of the sound wavefront in the lower medium,
(b) the pressure reflection and transmission coefficients, and
(c) the sound intensity reflection and power transmission coefficients.

Assume the salinity is 35 parts per thousand and neglect changes of density.

4.3. Plane acoustic waves propagating in air are incident normally upon a plane interface between the air and a second medium. A probe measuring acoustic intensity is inserted into the region in front of the interface and detects a standing-wave pattern the minima of which have 1 % of the intensity of the maxima. If the velocity of sound in air is 331 m/sec and the density of air is $1 \cdot 2$ kg/m³ determine the characteristic acoustic impedance of the second medium.

4.4. If the sea-floor, consisting of mud, is plane and at a constant depth of 10 m, and a sound source radiating at a frequency of 1 kc/s is situated in midwater, calculate

(a) the pressure amplitude reflection coefficient for normal incidence at both boundaries,
(b) the critical angle at both boundaries,
(c) the phase change occurring at the lower boundary when the angle of incidence is 70°,
(d) the initial angles of incidence of all rays which may propagate assuming the sea-bed to be rigid,
(e) the reflection coefficient of the wave incident upon the sea-floor at an angle of 70° and the attenuation coefficient of the resulting inhomogeneous wave if the attenuation coefficient of longitudinal waves in mud is 0·42 nepers/m.

Assume that the density and propagation velocity of mud are 2000 kg/m³ and 1670 m/sec respectively, and the corresponding figures for air are 1·21 kg/m³ and 343 m/sec respectively.

4.5. The velocity of sound in the sea is found to increase linearly from 1480 m/sec at the surface to 1500 m/sec at a depth of 200 m. It then decreases linearly with increase in depth becoming 1480 m/sec again at a depth of 2200 m. An omnidirectional transducer is situated at a depth of 50 m. Calculate

(a) the horizontal range to the point where the initially horizontal ray strikes the surface,
(b) the range of the point where the limiting ray becomes horizontal at the 200 m level,
(c) the initial angle of this limiting ray,
(d) the horizontal separation at a depth of 1700 m of the limiting ray and one which starts out with an initial angle of 5° greater (measured below the horizontal).

5.1. Determine the components of the equivalent circuit for a half-wave resonant air-backed quartz transducer of 4 cm² active area radiating into sea-water at a frequency of 2 Mc/s: $S_{11} = 1·15 \times 10^{-11}$ m²/newton, $e_{11} = 0·17$ coulomb/m², $\rho = 2·65 \times 10^3$ kg/m³, $\varepsilon = 4·45$, $\varepsilon_0 = 8·854 \times 10^{-12}$ F/m, $c = 5750$ m/s.

5.2. Determine the bandwidth of a half-wave resonant, lead zirconate-titanate transducer radiating into sea-water when it is air-backed, symmetrically loaded and operating with one face in contact with aluminium, if its thickness is 6×10^{-3} m, the velocity of propagation of sound in it is 2980 m/sec and its density is 7600 kg/m³.

5.3. The following values of conductance and susceptance were obtained from a test on a lead zirconate-titanate transducer. Determine the relative bandwidths in air and water and the radiation efficiency of the transducer.

In water

Frequency (kc/s)	Conductance (mmhos)	Susceptance (mmhos)	Frequency (kc/s)	Conductance (mmhos)	Susceptance (mmhos)
235·1	0·146	0·700	222·1	1·120	0·317
232·2	0·110	0·460	221·3	1·180	0·520
229·1	0·200	0·270	220·6	1·120	0·726
227·6	0·280	0·165	219·9	1·010	0·881
226·15	0·402	0·071	219·4	0·908	0·945
224·9	0·603	0·007	218·3	0·706	1·020
224·7	0·670	0·007	217·5	0·660	1·020
223·9	0·810	0·041	216·7	0·500	0·985
222·6	1·020	0·201	216·0	0·310	0·890

In air

Frequency (kc/s)	Conductance (mmhos)	Susceptance (mmhos)	Frequency (kc/s)	Conductance (mmhos)	Susceptance (mmhos)
229·20	0·49	0	222·70	11·4	1·64
224·03	0·799	−2·35	222·60	10·2	4·06
223·36	2·69	−4·26	222·35	9·18	5·18
223·34	4·05	−4·90	222·30	7·08	6·23
223·33	6·02	−5·10	222·25	5·12	6·23
223·22	8·08	−4·70	222·20	4·03	5·95
222·86	9·28	−4·00	222·10	3·07	5·46
222·83	10·50	−2·77	221·98	2·10	4·76
222·80	11·20	−1·47	221·52	1·06	3·48
222·75	11·50	0·01	217·88	0·11	1·22

5.4. A transducer consists of a half-wave electrostrictive element resonant at 500 kc/s which is bonded to a backing plate of stainless-steel, one-quarter wavelength in thickness, backed in turn by air. The transducer receives a signal from a 500 kc/s omnidirectional source radiating an acoustic power of 1 watt. If both source and transducer are immersed in sea-water and separated by a distance of 100 m, determine the r.m.s. open-circuit output voltage of the transducer. The g-coefficient, which is the electric field produced as a result of a unit applied stress under open circuit conditions, is for this particular element $28·3 \times 10^{-3}$ V m/newton The velocity of sound in the electrostrictive element is 2980 m/s.

5.5. The protective diaphragm of an oil-immersed transducer consists of a sheet of aluminium, $\lambda/20$ in thickness. Neglecting absorption losses, determine the power transmission coefficient of this diaphragm. Assume the characteristic impedances of the oil and sea water to be the same.

6.1. An electroacoustic transducer of dimensions small compared with the wavelength of the acoustic signal may be assumed to give an electrical output, from a received acoustic wave of given intensity, independent of the direction of arrival of the wave. Examine how you could use a number of such transducers in a given plane to determine the direction of a wave in that plane, and design an array of transducers which, at a wavelength of 10 cm, will enable the direction to be determined with an accuracy of the order of $\pm 10°$ free of major ambiguity within a sector of 90°.

6.2. Find the directional function of a continuous line array in which the sensitivity (on reception) varies from a maximum at the centre ($r = 0$) to a minimum at the ends ($r = \pm l/2$) according to the exponential-taper law:

$$T(r) = T_0 \exp\left(- a \,|\, r \,|\right).$$

Sketch and discuss the form of the directional functions for an array of length $l = 2$, and for $a = 1$ and $a = 3$, compare these functions with others you know, and discuss what would influence the choice of a suitable value of a for practical purposes.

6.3. A continuous line array, which may be divided into sections as required, is to be used for determining the direction of a received signal by the null method. It is required to have a directional pattern with a zero response on the axis normal to the line of the array, with a positive peak response at about 10° to one side of this axis and a negative peak response at the same angle on the other side. Devise a suitable method of obtaining this response and determine without detailed calculation the approximate length of the array in terms of the wavelength of the signal.

6.4. Using the process of synthesis involving the superposition of $(\sin x)/x$ patterns with different deflection, design a line array to give an approximately trapezoidal directional function in which the flat top extends over an angle twice the 3-dB beamwidth of the array when used with uniform sensitivity along its length, and in which the 3-dB beamwidth is four times that of the uniform array. Sketch the final directional pattern obtained.

6.5. Calculate the directivity factor of a line transducer array which has a directional function, expressed in terms of voltage response against physical angle θ, consisting of

  (a) an isosceles triangle of base width ± 0.2 rad,
  (b) a semicircle when θ is expressed in radians and the peak voltage response is unity.

In neither case are there any secondary responses.

Comment on the likelihood of being able to realize these patterns in practice.

6.6. Determine the directivity index (D.I.) and noise figure (N.F.) of a line array with the taper function

$$T(r) = (2/l) \cos (\pi r/l)$$

where r is the distance along the array measured from zero at the centre, and the length of the array is l. Assume l contains a large number of wavelengths. Discuss the relationship between D.I. and N.F.

6.7. If the dimensions of a rectangular transducer l_1, l_2 are large compared with the acoustic wavelength, show that the directivity factor D.F. is given by the expression

$$\text{D.F.} = \frac{4\pi l_1 l_2}{\lambda^2}$$

6.8. Design a sonar system (in outline) which will be capable of detecting a sphere of diameter 4 m at a range of 2000 m with an angular resolution of $1°$ and a range resolution of 5 m. It may be assumed that propagation is free of refraction, fluctuation and reverberation, but that the noise intensity level at the receiving transducer is -160 dB relative to 1 Wm^2 in a 1 c/s band.

Solutions

2.1. Resonant angular frequency $\omega_0 = 1/\sqrt{[L(C_1 + C_2)]}$.

Resistance of tuned circuit $L_1 C_2 = 100 \sqrt{L/C_2}$

$$= R', \text{ say.}$$

Let R_1, R_2 and R' in parallel be a resistance R.
Then 3-dB bandwidth in rad/sec $\simeq 1/R(C_1 + C_2)$.
Growth of voltage envelope across R_2, for final steady-state value of unity, is $1 - \exp(-at)$ where $a = 1/2R(C_1 + C_2)$.
Rate of rise at half the final amplitude $= a$

$$= 0 \cdot 5 \text{ times 3-dB bandwidth in rad/sec}$$
$$\simeq 1 \cdot 57 \text{ times 3-dB bandwidth in c/s.}$$

2.2. The correlation coefficient is

$$\psi = \frac{\overline{v_{N1} \cdot v_{N2}}}{V_{N1} \cdot V_{N2}} \text{ where } v_{N1} \text{ and } v_{N2} \text{ are given by eqns (2.11) and (2.12);}$$

now

$$v_{N1} \cdot v_{N2} = \left(a_0 \sum_r \cos \omega_r t + a_1 \sum_p \cos \omega_p t\right)\left(a_0 \sum_r \cos \omega_r t + a_2 \sum_q \cos \omega_q t\right)$$

$$= a_0 \left(\sum_r \cos \omega_r t\right)^2 + a_0 a_1 \sum_r \cos \omega_r t \sum_p \cos \omega_p t$$

$$+ a_0 a_2 \sum_r \cos \omega_r t \sum_q \cos \omega_q t + a_1 a_2 \sum_p \cos \omega_p t \sum_q \cos \omega_q t$$

so

$$\overline{v_{N1} \cdot v_{N2}} = a_0^2 \cdot \tfrac{1}{2}n.$$

Now

$V_{N1} = V_{N2}$ and $a_1 = a_2$ in the given problem and therefore

$$V_{N1} \cdot V_{N2} = V_{N1}^2 = a_0^2 \cdot \tfrac{1}{2}n + a_1^2 \cdot \tfrac{1}{2}m$$
$$= (a_0^2 + a_1^2)\tfrac{1}{2}n \text{ if we put } n = m.$$

Therefore

$$\psi = a_0^2/(a_0^2 + a_1^2).$$

If $\psi = 0\cdot5$ as in the problem, then $a_0 = a_1$, where the water noise is represented by v_{N1} and v_{N2} at the two hydrophones. The noise introduced after the hydrophones is of r.m.s. voltage (referred to the hydrophones) equal to V_{N1} but is uncorrelated both between channels and, of course, with the water noise. Thus when the channels are combined together, we have mean-square voltages (referred to the hydrophones) thus:

$\tfrac{1}{2}V_{N1}^2$ in each channel correlated;

$\tfrac{3}{2}V_{N1}^2$ in each channel uncorrelated.

The total mean-square noise voltage is therefore

$$2\,V_{N1}^2 + 3\,V_{N1}^2 = 5\,V_{N1}^2$$

and the signal voltage is doubled. The r.m.s. signal–noise ratio at the output is therefore $2/\sqrt{5}$ times that at the hydrophones.

2.3. Let $S = $ Sum of random quantities
i.e.

$$S = a + b + c + d + e \ldots, \text{ etc.}$$

1st sample

$$S_1 = a_1 + b_1 + c_1 + d_1 + e_1 \ldots$$

2nd sample

$$S_2 = a_2 + b_2 + c_2 + d_2 + e_2 \ldots$$

nth sample

$$S_n = a_n + b_n + c_n + d_n + e_n \ldots$$

Mean $\quad S = (S_1 + S_2 + \ldots S_n) \times \dfrac{1}{n}$ as $n \to \infty$

$$= \left\{ \begin{array}{l} a_1 + a_2 + a_3 + a_4 \ldots a_n \\[4pt] + b_1 + b_2 + b_3 \quad \ldots b_n \\[4pt] + c_1 + c_2 + c_3 \quad \ldots c_n \\[4pt] + d_1 + d_2 + d_3 \quad \ldots d_n \\[4pt] \ldots, \text{ etc.} \end{array} \right\} \times \dfrac{1}{n}$$

$$= \dfrac{a_1 + \ldots a_n}{n} + \dfrac{b_1 + \ldots b_n}{n} + \dfrac{c_1 + c_2 \ldots c_n}{n}, \text{ etc.}$$

i.e.

$$\bar{S} = \bar{a} + \bar{b} + \bar{c} + \bar{d} + \bar{e}, \text{ etc.}$$

2.4. Since V_A and V_B have the same r.m.s. values, the correlation coefficient may be written

$$\psi = \frac{\overline{(V_A + V_B)^2} - 2\,\overline{V_A^2}}{2\,\overline{V_A^2}}.$$

Let the source voltage of the rth source be written

$$V_r = \sum_q v_{rq} \cos(\omega_q t + a_q).$$

The sum $(V_A + V_B)$ contains two series of terms from this source:

$$\Sigma\, v_{rq} \cos[\omega_q t + a_q + (m - r)\,\phi]$$

and

$$\Sigma\, v_{rq} \cos[\omega_q t + a_q + (m + r)\,\phi]$$

which combine to give

$$2 \cos r\phi \,.\, \Sigma\, v_{rq} \cos(\omega_q t + a_q + m\phi)$$

which may be written as

$$2 \cos r\phi \,.\, V_r'$$

where V_r' has the same mean-square value as V_r.

Therefore, counting all sources,

$$V_A + V_B = \sum_{r=-m}^{r=+m} 2 \cos r\phi \,.\, V_r'$$

and since the mean-square values of the noise sources are all the same (and equal to V_N^2 say),

$$\overline{(V_A + V_B)^2} = V_N \sum_{r=-m}^{r=+m} (2 \cos r\phi)^2.$$

Also

$$\overline{V_A^2} = \overline{V_B^2} = nV_N.$$

Thus

$$\psi = \frac{1}{n} \sum_{r=-m}^{r=+m} \cos 2r\phi = \frac{\sin n\phi}{n \sin \phi}.$$

This is zero when $\phi = \pm\, \pi/n,\ \pm\, 2\pi/n$, etc., which is the required answer.

3.1. Let p_x and $p_x + \delta p_x$ be the shear stresses at x and $x + \delta x$ respectively. Now $p\alpha$-rate of deformation, and therefore

$$p_x = -\mu \frac{\partial}{\partial t}\left(\frac{\partial \xi_s}{\partial x}\right)$$

$$p_x + \delta p_x = -\left[\mu \frac{\partial}{\partial t}\left(\frac{\partial \xi_s}{\partial x}\right) + \frac{\partial}{\partial x}\left(\mu \frac{\partial^2 \xi_s}{\partial x \partial t}\right)\delta x\right].$$

Resultant stress $= (p_x + \delta p_x) - p_x = \dfrac{\mu\, \partial^3 \xi_s}{\partial x^2 \partial t}\,\delta x.$

Now using Newton's second law

$$\delta p_x = \rho \delta x \frac{\partial^2 \xi_s}{\partial t^2} = \mu \frac{\partial^3 \xi_s}{\partial x^2 \partial t}.$$

The wave equation is therefore

$$\frac{\partial^2 \xi_s}{\partial t^2} = \frac{\mu}{\rho} \frac{\partial^3 \xi_s}{\partial x^2 \partial t}. \tag{1}$$

Assume a solution of the form

$$\xi_s = \hat{\xi}_s \exp [j(\omega t - k'x)].$$

Then eqn. (1) becomes

$$- j\omega\rho = \mu(k')^2$$

which can be written in the form

$$k' = \pm \sqrt{\frac{\omega\rho}{2\mu}} (1 + j)$$

i.e.

$$k' = k + jk \text{ where } k = \pm \sqrt{\frac{\omega\rho}{2\mu}}.$$

The solution of the wave equation is therefore

$$\xi_s = \hat{\xi} \exp(-kx) \exp [j(\omega t - kx)]$$

where

$$k = a_s = \text{shear attenuation coefficient}$$

$$= \sqrt{\frac{\omega\rho}{2\mu}}.$$

Now

$$\omega = 4\pi \times 10^6 \text{ rad/sec}$$

$$\mu = 10^{-3} \text{ newton sec/m}^2$$

$$\rho = 998 \text{ kg/m}^3$$

therefore

$$a_s = 2 \cdot 5 \times 10^6 \text{ nepers/m}.$$

3.2. At 2 kc/s,

$$k = \frac{\omega}{c} = 8 \cdot 4$$

The phase difference θ between the acoustic pressure and the particle velocity is given by

$$\theta = \tan^{-1}(1/kr).$$

At 10 cm,

$$kr = 0 \cdot 84 \quad \theta = 49° 58'.$$

At 1 m,

$$kr = 8 \cdot 4 \quad \theta = 6° 45'.$$

At 10 m,

$$kr = 84 \quad \theta = 0° 40'.$$

The specific acoustic impedances are given by

$$Z = Z_0 \frac{jkr}{1 + jkr} = Z_0 \left(\frac{k^2r^2}{1 + k^2r^2} + j \frac{kr}{1 + k^2r^2} \right).$$

At 10 cm,

$$Z = Z_0 \frac{0 \cdot 84j}{1 + 0 \cdot 84j} = Z_0 (0 \cdot 41 + j \, 0 \cdot 494)$$

$$= 6 \cdot 15 \times 10^5 + j \, 7 \cdot 4 \times 10^5 \, \Omega.$$

At 1 m,

$$Z = Z_0 \frac{8 \cdot 4j}{1 + 8 \cdot 4j} = Z_0 (0 \cdot 99 + j \, 0 \cdot 118)$$

$$= 1 \cdot 49 \times 10^6 + j \, 1 \cdot 77 \times 10^5 \, \Omega.$$

At 10 m,

$$Z = Z_0 \frac{84j}{1 + 84j} = Z_0 (1 + j \, 0 \cdot 012)$$

$$= 1 \cdot 5 \times 10^6 + j \, 1 \cdot 8 \times 10^4 \, \Omega.$$

3.3. The acoustic pressure of a spherical wave is given by

$$p = \frac{A}{r} \exp [j(\omega t - kr)]$$

but

$$u = -\frac{1}{j\omega\rho_0} \frac{\partial p}{\partial x}$$

$$= \frac{1}{j\omega\rho_0} \cdot \frac{A}{r} \left(\frac{1}{r} + jk \right) \exp [j(\omega t - kr)]$$

$$= \frac{p}{j\omega\rho_0} \left(\frac{1}{r} + jk \right)$$

Thus

$$Y = \frac{p}{u} = \frac{(1/r) + jk}{j\omega\rho_0} = Y_0 \frac{1 + jkr}{jkr}.$$

Rearranging into real and imaginary parts

$$Y = Y_0(1 - j/kr).$$

Thus the equivalent circuit becomes as shown in Fig. P3.

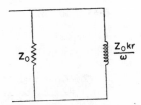

FIG. P3. Equivalent circuit of medium.

The acoustic intensity at 10 m from the source $= \dfrac{W}{4\pi r^2} = 0\cdot8$ W/m^2.

The peak acoustic pressure, given by

$$I = \hat{p}^2/2\rho_0 c$$

equals

$$1\cdot55 \times 10^3 \text{ newton/m}^2.$$

The acoustic pressure, p, and the particle velocity, u, are related by

$$p = \rho_0 c u \cos\theta \text{ where } \cos\theta = kr/\sqrt{(1 + k^2 r^2)}.$$

At 25 c/s and 10 m,

$$kr = \pi/3$$

therefore

$$\cos\theta = 0\cdot72,$$

and thus

$$\hat{u} = \frac{\hat{p}}{\rho_0 c \cos\theta} = 1\cdot44 \times 10^{-3} \text{ m/sec}.$$

But the particle displacement, ξ, velocity, u, and acceleration, a, are related by the expression, considering magnitudes only.

$$a = \omega u = \omega^2 \xi$$

therefore

$$\hat{\xi} = 9\cdot2 \times 10^{-5} \text{ m}$$

and

$$\hat{a} = 0\cdot216 \text{ m/sec}^2.$$

At the surface of the sphere the intensity is given by

$$W/4\pi = 80 \text{ W/m}^2.$$

The corresponding peak acoustic pressure is therefore $= 1\cdot55 \times 10^4$ newton/m^2.

At 25 c/s and 1 m,

$$kr = \pi/30$$

therefore

$$\cos\theta \simeq \pi/30$$

thus

$$\hat{u} = \frac{\hat{p}}{\rho_0 c \cos\theta} = 0\cdot1 \text{ m/sec};$$

therefore

$$\hat{\xi} = 0\cdot65 \times 10^{-3} \text{ m}.$$

and

$$\hat{a} = 15\cdot7 \text{ m/sec}^2.$$

3.4. Equation (3.38) states that

$$\mathbf{Z}_s = \mathbf{Z}_0 \frac{(\mathbf{Z}_T/\mathbf{Z}_0) + j \tan kl}{1 + j(\mathbf{Z}_T/\mathbf{Z}_0) \tan kl}$$

when

$$l = \lambda/4$$

this reduces to

$$\mathbf{Z}_s = \mathbf{Z}_0^2/\mathbf{Z}_T.$$

In the present case

$$\mathbf{Z}_0 = 47 \cdot 0 \times 10^6 \text{ kg/m}^2 \text{ sec},$$
$$\mathbf{Z}_T = 1 \cdot 5 \times 10^6 \text{ kg/m}^2 \text{ sec}$$

therefore

$$\mathbf{Z}_s = 1 \cdot 46 \times 10^8 \text{ kg/m}^2 \text{ sec}.$$

If l is increased to $\lambda/2$, $kl = \pi$, $\tan kl = 0$

$$\mathbf{Z}_s = \mathbf{Z}_0 \frac{\mathbf{Z}_T}{\mathbf{Z}_0} = \mathbf{Z}_T$$

The impedance is the same as if the $\lambda/2$ element were omitted altogether.

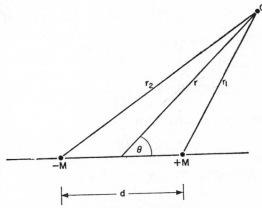

Fig. P4. An acoustic doublet source.

3.5. (See Fig. P4.) The acoustic pressure due to a point source of strength M at a distance r is given by

$$p_r = \mathrm{j} \frac{\rho_0 \omega M}{4\pi r} \exp [\mathrm{j}(\omega t - kr)].$$

Therefore the acoustic pressure at point O due to both sources is given by

$$p_0 = \mathrm{j} \frac{\rho_0 \omega M}{4\pi r_1} \exp [\mathrm{j}(\omega t - kr_1)] - \mathrm{j} \frac{\rho_0 \omega M}{4\pi r_2} \exp [\mathrm{j}(\omega t - kr_2)]$$

$$= \mathrm{j} \frac{\rho_0 \omega M}{4\pi} \left[\frac{\exp(-\mathrm{j}kr_1)}{r_1} - \frac{\exp(-\mathrm{j}kr_2)}{r_2} \right] \exp(\mathrm{j}\omega t)$$

where, since

$$r \gg d, \ r_1 \simeq r - (d/2) \cos \theta \text{ and } r_2 \simeq r + (d/2) \cos \theta.$$

Therefore

$$p_0 = j \frac{\rho_0 \omega M}{4\pi} \left\{ \frac{\exp[-jk(r - d/2 \cdot \cos \theta)]}{(r - d/2 \cdot \cos \theta)} \right.$$

$$\left. - \frac{\exp[-jk(r + d/2 \cdot \cos \theta)]}{(r + d/2 \cdot \cos \theta)} \right\} \exp(j\omega t)$$

$$= \frac{j\rho_0 \omega M \exp(-jkr)}{4\pi(r^2 - d^2/4 \cdot \cos^2 \theta)} [d \cos \theta \cos(kd/2 \cdot \cos \theta)$$

$$+ j2r \sin(kd/2 \cdot \text{os } \theta)] \exp(j\omega t).$$

Since d is small kd will be small also; therefore

$$\sin(kd/2 \cdot \cos \theta) \simeq kd/2 \cdot \cos \theta, \cos(kd/2 \cdot \cos \theta) \simeq 1$$

and

$$d^2/4 \cdot \cos^2 \theta \simeq 0$$

therefore

$$p_0 = \frac{j\rho_0 \omega M \exp[j(\omega t - kr)]}{4\pi r^2} (d \cos \theta + 2jr \, kd/2 \cdot \cos \theta)$$

$$\frac{j\rho_0 \omega M \cdot d \cos \theta}{4\pi r^2} (1 + j \, kr) \exp[j(\omega t \cdot kr)].$$

Now $kr \gg 1$; therefore

$$p_0 = \frac{j\rho_0 \omega Md \cos \theta}{4\pi r^2} \cdot j \, kr \exp[j(\omega t - kr)].$$

The magnitude of the acoustic pressure is therefore

$$p_0 = \frac{\rho_0 c k^2 Md \cos \theta}{4\pi r},$$

since $\omega = kc$.

The radiated pressure pattern is shown in Fig. P5.

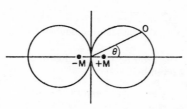

Fig. P5. Pressure pattern of an acoustic doublet.

4.1. Signals travelling via the surface reflection and the direct path interfere with one another. The geometry is shown in Fig. P6.

$$\text{Path TR} = R \left[1 + \left(\frac{h_2 - h_1}{R} \right)^2 \right]^{1/2}$$

$$= R \left[1 + \frac{1}{2} \left(\frac{h_2 - h_1}{R} \right)^2 \right].$$

Path TOR $= R \left[1 + \left(\dfrac{h_2 + h_1}{R} \right)^2 \right]^{1/2}$

$$= R \left[1 + \frac{1}{2} \left(\frac{h_2 + h_1}{R} \right)^2 \right].$$

The path length difference $= \text{TOR} - \text{TR} = \dfrac{2h_1h_2}{R}.$

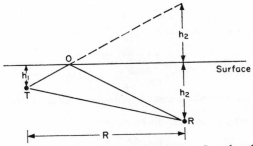

FIG. P6. Geometry of direct and surface-reflected paths.

If this path difference is $n\lambda/2$ then destructive interference occurs. Thus

$$\frac{2h_1h_2}{R} = \frac{n\lambda}{2}, \text{ i.e. } R = \frac{4h_1h_2}{n\lambda}.$$

At 20 kc/s,
$$\lambda = 0 \cdot 075 \text{ m}.$$

Taking $n = 1$ for last minimum, neglecting phase change at surface:

$$R = 10{,}667 \text{ m}.$$

4.2. The velocity of propagation at 60 m in the upper layer $= 1506$ m/sec. The velocity of propagation at 60 m in the lower layer $= 1471$ m/sec. On either side of the boundary assume the density remains constant; using Snell's Law, the angle of the ray in the lower layer is given by

$$\sin \theta_2 = \frac{\sin \theta_1 c_2}{c_1},$$

there $\theta_2 = 43° \, 42'$. The wavefront being normal to the ray makes an angle of $46° \, 18'$ with the horizontal.

The pressure amplitude reflection and transmission coefficients, V and W, are given by

$$V = \frac{Z_2 - Z_1}{Z_2 + Z_1}, \quad W = \frac{2Z_2}{Z_2 + Z_1}$$

where

$$Z_1 = \frac{\rho_1 c_1}{\cos \theta_1} = 2129 \, \rho_1 \text{ ohms}$$

and

$$Z_2 = \frac{\rho_1 c_2}{\cos \theta_2} = 2036 \, \rho_1 \text{ ohms}.$$

Therefore

$$V = -2 \cdot 23 \times 10^{-2}$$

and

$$W = 0 \cdot 9777.$$

The sound intensity reflection and power transmission coefficients a_r, a_t' are given by

$$a_r = V^2 = 5 \times 10^{-4},$$

$$a_t' = \frac{4Z_1 Z_2}{(Z_2 + Z_1)^2} = 0 \cdot 9995.$$

4.3. Let the characteristic acoustic impedances of the two media be Z_1 and Z_2 respectively.

The reflection coefficient at the boundary

$$V = \frac{Z_2 - Z_1}{Z_2 + Z_1}.$$

The equation of the particle amplitude is thus

$$\xi = \hat{\xi} \sin(\omega t - kx) + V\hat{\xi} \sin(\omega t + kx)$$

$$= \hat{\xi}(1 + V) \sin \omega t \cos kx - \hat{\xi}(1 - V) \cos \omega t \sin kx.$$

This represents two waves

      (a) max. amp. $\hat{\xi}(1 + V)$ which is zero when $\cos kx = 0$

      (b) max. amp. $\hat{\xi}(1 - V)$ which is zero when $\sin kx = 0$

The resultant of these two waves is a standing wave pattern which has nodes and antinodes of amplitude $\xi(1 - V)$, $\xi(1 + V)$ respectively. Therefore

$$\frac{\text{Max. amplitude}}{\text{Min. amplitude}} = \frac{1 + V}{1 - V} = \frac{\xi_{max}}{\xi_{min}}$$

i.e.

$$V = \frac{\xi_{max} - \xi_{min}}{\xi_{max} + \xi_{min}}.$$

The probe measures intensity which is proportional to (displacement)^2. Thus if

$$\xi_{max} = 1,$$

$$\xi_{min} = 0 \cdot 1,$$

therefore

$$V = \frac{9}{11} = \frac{Z_2 - Z_1}{Z_2 + Z_1};$$

now
$$Z_1 = \rho_1 c_1 = 400 \text{ kg/m}^2 \text{ sec};$$

therefore
$$Z_2 = 4000 \text{ kg/m}^2 \text{ sec}.$$

4.4.

(a) The pressure amplitude reflection coefficient at normal incidence at the upper boundary is given by
$$V = \frac{R_2 - R_1}{R_2 + R_1}.$$

At the upper boundary
$$V \simeq 1.$$

At the lower boundary
$$V = 0 \cdot 37.$$

(b) The critical angle, θ_c, is given by
$$\theta_c = \text{does not exist}$$

At the upper boundary
$$\theta_c = 13° \ 18'.$$

At the lower boundary
$$\theta_c = 64° \ 10'.$$

(c) The phase change at a boundary is given by
$$\phi = - 2 \tan^{-1} \frac{\sqrt{(\sin^2 \theta_i - \eta^2)}}{b \cos \theta_i}.$$

At the lower boundary
$$\phi = 42° \ 24'.$$

(d) The equation defining the initial angles of the tenable ray paths when the sea-floor is rigid is
$$\cos \theta = (n - \tfrac{1}{2}) \ \lambda/2d \text{ where } d \text{ is the depth.}$$

Thus the initial angles of all the tenable paths are given by
$$\theta = \cos^{-1} (n - \tfrac{1}{2}) \ (0 \cdot 075) \text{ where } n = 1, 2, 3, \text{ etc.}$$

i.e.
$$\theta = 88°, 83 \cdot 5°, 79°, 75°, 70°, 65 \cdot 5°, 61°, 56°, 50°, 44 \cdot 5°, 38°, 30°, 20°, \text{ etc.}$$

(e) The intensity reflection coefficient of a wave incident upon an absorbing medium is given by
$$\alpha_r = V^2 = \frac{(b \cos \theta_i + g)^2 + h^2}{(b \cos \theta_i - g)^2 + h^2}$$

where
$$b = \rho_M/\rho_W,$$
$$g = - (1/\sqrt{2})[\sqrt{(A^2 + B^2)} - A]^{1/2},$$
$$h = (1/\sqrt{2})[A + \sqrt{(A^2 + B^2)}]^{1/2},$$
$$A = \sin^2 \theta_i - \eta_0^2,$$
$$B = 2\eta_0^2 \alpha,$$
$$\alpha = a/k.$$

At 1 kc/sec,

$$k = 3 \cdot 7$$

therefore

$$a = -0 \cdot 114,$$
$$A = 0 \cdot 073,$$
$$B = -0 \cdot 183,$$
$$g = -0 \cdot 25,$$
$$h = 0 \cdot 37,$$

hence

$$a_r = 0 \cdot 32.$$

The attenuation coefficient a_x of the resulting inhomogeneous wave is given by

$$a_x = k_1 \sqrt{(\sin^2 \theta_i - \eta^2)}$$
$$= 1 \cdot 11 \text{ nepers/m.}$$

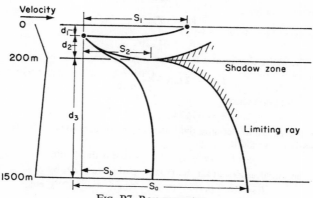

FIG. P7. Ray geometry.

4.5. (See Fig. P7.) There are clearly two regions, with velocity gradients

$$g_1 = \frac{1500 - 1480}{200} = 0 \cdot 1 \text{ m/sec m.}$$

$$g_2 = \frac{1480 - 1500}{2200 - 200} = -0 \cdot 01 \text{ m/sec m.}$$

The velocity of propagation at the transducer

$$c_{50} = 1480 + 50 (0 \cdot 1) = 1485 \text{ m/sec.}$$

The radius of curvature of the initially horizontal ray,

$$R_{50}(1) = \frac{C_{50}}{g_1} = 14,850 \text{ m.}$$

The range to the point where the initially horizontal ray strikes the surface is

$$S_1 = \sqrt{(2\ R_{50}(1)\ d_1)} = 1240\ \text{m}.$$

The velocity of propagation at the interface between the two regions

$$c_{200} = 1500\ \text{m/sec}.$$

The radius of curvature of the limiting ray

$$R_{200}(1) = \frac{c_{200}}{g_1} = 15{,}000\ \text{m}.$$

The range to the point where the limiting ray becomes tangential to the 200 m level,

$$S_2 = \sqrt{(2\ R_{200}(1)\ d_2)} = 2120\ \text{m}.$$

The initial angle ϕ of the limiting ray is given by eqn. (4.48)

$$\cos \phi_1 = 1 - \frac{d_2}{R_{200}(1)} = 0{\cdot}99$$

therefore

$$\theta_1 = 8°\ 6'\ \text{below the horizontal}.$$

The radius of curvature of the limiting ray in region 2 is

$$R_{200}(2) = \frac{c_{200}}{g_2} = 150{,}000\text{m}.$$

The angle of the limiting ray at a depth of 1700 m which corresponds to $d_3 = -\ 1500$ m is given by

$$\cos \phi_l = 1 - \frac{d_3}{R_{200}(2)} = 0{\cdot}99.$$

$$\phi_l = 8°\ 6'$$

Therefore the horizontal range at the 1700 m level is given by

$$S_a = S_2 + R_{200}(2)\ (\sin 0 - \sin 8°6') = 23{,}120\ \text{m}.$$

The initial angle of the second ray to pass into region 2 is

$$8°\ 6' + 5° = 13°\ 6'.$$

Its radius of curvature given by

$$R'_{200}{}^{(1)} = cp/g_1 \cos \theta p = -\ 152000\text{m}.$$

and its angle of inclination at the boundary is given by eqn. (4.49) is $10°\ 18'$; it will enter region 2 with this angle at a distance of $S_3 = 730$m.

The radius of curvature in region 2

$$R'_{200}{}^{(2)} = cp/g'_2 \cos \theta p = -\ 15200\text{m},$$

and the angle of this ray at a depth of 1700m which once again corresponds to $d_3 = -\ 1500$, as given by eqn (4.49) is $13°\ 6'$.

The horizontal range at the 1700 m level is therefore given by

$$S_b = S_3 + R'_{200}{}^{(2)}\ (\sin 10°\ 18' - \sin 13°6') = 8030\ \text{m}.$$

Thus the horizontal separation at a depth of 1700 m is

$$S_a - S_b = 15,100 \text{ m.}$$

5.1. The thickness of the crystal

$$l = \lambda/2 = c/2f = 14 \cdot 4 \times 10^{-4} \text{ m.}$$

The transformation ratio

$$a = Ae_{11}/l = 0 \cdot 047 \text{ coulomb/m.}$$

The radiation impedance

$$R_R = \rho_0 c_0 A = 6 \times 10^2 \text{ kg/sec.}$$

The equivalent resistance

$$R = R_R/4a^2 = 6 \cdot 80 \text{ K}\Omega.$$

The equivalent inductance

$$L = m/8a^2 = 86 \text{ mH.}$$

The equivalent capacitance

$$c = \frac{8a^2 sl}{\pi^2 A} = 0 \cdot 073 \ \mu\mu\text{F.}$$

he static capacitance

$$C_0 = A\varepsilon/l = 11 \cdot 0 \ \mu\mu\text{F.}$$

The relevant equivalent circuit is shown in Fig. P8.

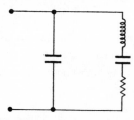

Fig. P8. Equivalent circuit of quartz transducer.

5.2. The transducer is a half-wave resonant element so its resonant frequency

$$= c/2l = 248 \text{ kc/s.}$$

Assuming the mechanical Q is the limiting factor:
(a) For air backed operation

$$Q = \pi/2 \ \frac{\rho c}{\rho_w c_w} = 23 \cdot 5.$$

The bandwidth $= f_0/Q = 10 \cdot 5$ kc/s.

(b) For symmetrically loaded operation

$$Q = \pi/2 \, \frac{\rho c}{2\rho_w c_w} = 11\cdot 75.$$

The bandwidth $= 20\cdot 5$ kc/s.

(c) For aluminium-backed operation

$$Q = \pi/2 \, \frac{\rho c}{\rho_{Al} c_{Al} + \rho_w c_w} = 1\cdot 93.$$

The bandwidth $= 128$ kc/s.

In practice the electrical Q-factor would be the controlling factor in case (c).

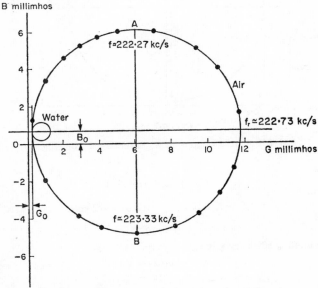

Fig. P9. Circle diagrams in air and water.

5.3. The circle diagrams in air and water are as shown in Figs. P9 and P10. From the circle diagram it can be seen that the static capacitance

$$C_0 = B_0/\omega$$
$$\simeq 370 \, \mu\mu\text{F}.$$

The 3-dB bandwidth is given by the difference in the frequencies across the diameter AB and $A'B'$ of the circle diagrams, thus in air

$$f_2 - f_1 = 223\cdot 33 - 222\cdot 27 = 1\cdot 06 \text{ kc/s};$$

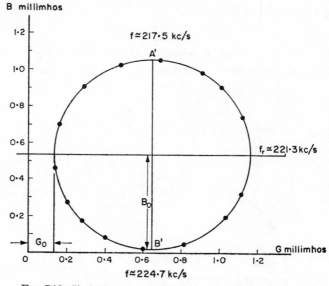

FIG. P10. Circle diagram in water (enlarged from Fig. P9).

hence

In water

$$Q = f_r/f_2 - f_1 = 214.$$

hence

$$f_2 - f_1 = 224 \cdot 7 - 217 \cdot 5 = 7 \cdot 2 \text{ kc/s};$$

$$Q = 30 \cdot 7.$$

The radiating efficiency η is given by eqn. (5.52) as

$$\eta = \frac{d_2(d_1 - d_2)}{d_1(a + d_2)}$$

where

$$d_1 = \frac{1}{R_L} = \text{diameter of circle in air}$$

$$= 11 \text{ mmhos.,}$$

$$d_2 = \frac{1}{R_L + R_R} = \text{diameter of circle in water}$$

$$= 1 \cdot 08 \text{ mmhos.,}$$

and

$$a = \frac{1}{R_D} = G_0 = 0 \cdot 1 \text{ mmhos.}$$

Therefore

$$\eta = 83\%.$$

5.4. Acoustic intensity at a distance of 100 m from source is

$$I = \frac{1}{4\pi r^2} = 8 \times 10^{-6} \text{ W/m}^2.$$

The acoustic pressure p is given by

$$I = \frac{\hat{p}}{2\rho_0 c_0} = \frac{p_{\text{rms}}^2}{\rho_0 c_0}.$$

Therefore

$$p_{\text{rms}} = 3 \cdot 46 \text{ newton/m}^2.$$

The impedance Z_S presented by a quarter-wave transformer is $Z_S = Z_0^2/Z_T$ where Z_T is the impedance of the medium in contact with the opposite face of the transformer. Since $Z_0 \gg Z_T$ the backing can be considered as rigid.

The total r.m.s. stress T acting on the element is $3 \cdot 46$ newton/m^2 so the resulting electric field across the element given by

$$E = gT$$
$$= 96 \text{ mV/m}.$$

Now

$$c = 2980 \text{ m/sec}, \qquad f = 500 \text{ kc/s}, \qquad \lambda = 6 \times 10^{-3} \text{ m}.$$

The thickness of the element is

$$\lambda/2 = 3 \times 10^{-3} \text{ m}.$$

Now

$$V = El;$$

therefore the r.m.s. open circuit voltage $= 288 \ \mu\text{V}$.

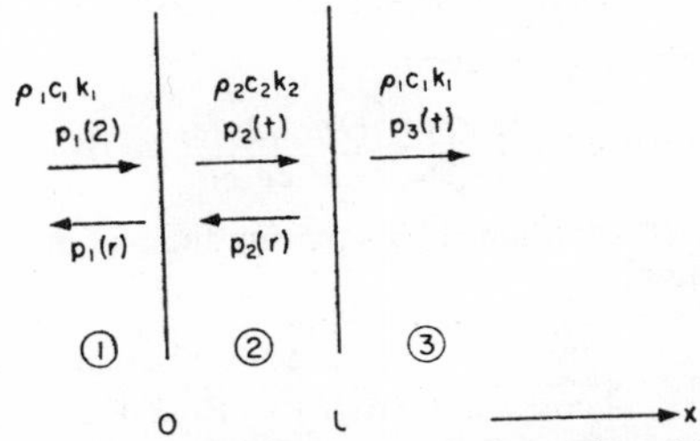

FIG. P11. Transmission of plane waves through a thin diaphragm.

5.5. Referring to Fig. P11,

$$p_1(i) = \hat{p} \exp [j(\omega t - k_1 x)]. \tag{1}$$
$$p_1(r) = A\hat{p} \exp [j(\omega t + k_1 x)]. \tag{2}$$
$$p_2(t) = B\hat{p} \exp [j(\omega t - k_2 x)]. \tag{3}$$

$$p_2(r) = C\hat{p} \exp [j(\omega t + k_2 x)]. \tag{4}$$

$$p_3(t) = D\hat{p} \exp [j(\omega t - k_1(l - x))]. \tag{5}$$

Using the boundary conditions

(1) Pressures on either side of the boundary must be equal at $x = 0$ and $x = l$.

$$\hat{p}(1 + A) = \hat{p}(B + C). \tag{6}$$

$$\hat{p}[B \exp (-jk_2 l) + C \exp (jk_2 l)] = \hat{p}D. \tag{7}$$

(2) Particle velocities on either side of the boundary constant at $x = 0$ and $x = l$.

$$\frac{\hat{p}}{\rho_1 c_1} - \frac{A\hat{p}}{\rho_1 c_1} = \frac{B\hat{p}}{\rho_2 c_2} - \frac{C\hat{p}}{\rho_2 c_2}. \tag{8}$$

$$\frac{\hat{p}B \exp (-jk_2 l)}{\rho_2 c_2} - \frac{\hat{p}C \exp (jk_2 l)}{\rho_2 c_2} = \frac{\hat{p}D}{\rho_1 c_1}. \tag{9}$$

Solve to find D. Let

$$\rho_1 c_1 = R_1, \qquad \rho_2 c_2 = R_2.$$

From (6) and (8):

$$2R_2 = (R_2 + R_1) B + (R_2 - R_1) C. \tag{10}$$

From (7) and (9):

$$B = D \frac{(R_1 + R_2)}{2R_1} \exp (jk_2 l). \tag{11}$$

$$C = D \frac{(R_1 - R_2)}{2R_1} \exp (-jk_2 l). \tag{12}$$

Substitution into eqn. (10) gives

$$4R_1 R_2 = D[(R_1 + R_2)^2 \exp (jk_2 l) + (R_2 - R_1)(R_1 - R_2) \exp (-jk_2 l)],$$

i.e.

$$D = \frac{4}{4 \cos k_2 l + j[2(R_1^2 + R_2^2)/R_1 R_2] \sin k_2 l}. \tag{13}$$

The power transmission coefficient

$$a_t' = \frac{\hat{p}_3(t)}{\hat{p}_1(i)} = \frac{(D\hat{p})^2/2\rho_1 c_1}{\hat{p}^2/2\rho_1 c_1} = D^2.$$

The magnitude of D^2 is obtained by squaring the real and imaginary parts of eqn. (13) and adding,

Therefore

$$a_T = \frac{4}{4 \cos^2 k_2 l + (R_2/R_1 + R_1/R_2)^2 \sin^2 k_2 l}.$$

Now

$$k_2 = 2\pi/\lambda_2 \text{ and } l = \lambda/20 \text{ thus } k_2 l = \pi/10$$

also

$$\rho_1 c_1 = 1 \cdot 5 \times 10^6 \text{ kg/m}^2 \text{ sec},$$

$$\rho_2 c_2 = 17 \times 10^6 \text{ kg/m}^2 \text{ sec},$$

therefore

$$a_T = 0 \cdot 82.$$

6.1. Fig. 6.5 shows that the sector free of major ambiguity may be taken as having a width from $- \lambda/2d$ to $+ \lambda/2d$ on the scale of sin θ. Therefore

$$10/2d = \sin 45° = 1/\sqrt{2}.$$

Therefore

$$d = 5\sqrt{2} \text{ cm} \simeq 7 \text{ cm}.$$

The accuracy of $\pm 10°$ suggests a beamwidth of 20° between zeros, i.e.

$$\frac{\lambda}{nd} = \frac{10}{5\sqrt{2}n} \simeq \sin 10° = 0 \cdot 174.$$

so that

$$n \simeq 2/(0 \cdot 174\sqrt{2}) \simeq 8 \cdot 1.$$

Thus an array of 8 elements at a spacing of 7 cm will meet the requirement.

6.2. $T(r) = T_0 \exp{(- a \mid r \mid)}.$

$$D(K) = \int_{-\infty}^{\infty} T(r) \cdot \exp{(- jKr)} \cdot dr$$

$$= 2T_0 \int_0^{l/2} \exp{(- ar)} \cos Kr \cdot dr$$

since the pattern is symmetrical

$$= T_0 \int_0^{l/2} \{\exp{[- (a + jK)]} r + \exp{[- (a - jK) r]}\} dr$$

$$= \frac{T_0}{a + jK} \{1 - \exp{[- (a + jK) l/2]}\}$$

$$+ \frac{T_0}{a - jK} \{1 - \exp{[- (a - jK) l/2]}\}$$

$$= \frac{2T_0}{a^2 + K^2} [a - a \exp{(- al/2)} \cos Kl/2$$

$$+ K \exp{(- al/2)} \sin Kl/2].$$

This is plotted for an array of length 2 units (i.e. $l/2 = 1$) in Fig. P12 for the two values of a.

The choice of a would depend on the specification regarding sidelobes or secondary responses, but clearly $a = 1$ gives a more generally satisfactory pattern than $a = 3$.

6.3. A suitable method is to divide the array into halves, reverse the phase of one output, and then add the outputs. This gives a directional pattern as given by eqn. (6.41) and Fig. 6.21. Zeros occur at $x = - 2\pi,\ 0,\ + 2\pi$ and so peaks must occur roughly at $x = - \pi$ and $+ \pi$. Thus

$$(\pi l/\lambda) \sin 10° \simeq \pi$$

giving

$$l/\lambda \simeq 5.$$

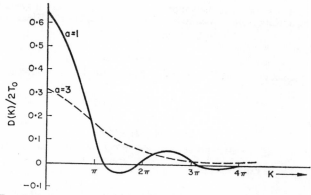

FIG. P12. Directional functions for exponential taper with $a = 1$ and $a = 3$.

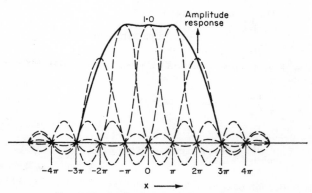

FIG. P13. Trapezoidal directional function.

6.4. The directional pattern is shown in Fig. P13, where the dashed curves show five component patterns and the full-line curve shows the resultant pattern. The taper function required is

$$1 + 2 \cos (s/m) \pi + (2/\sqrt{2}) \cos 2(s/m) \pi.$$

6.5.

(a) $D(\theta) = 1 - 5 \mid \theta \mid$ over the range $\theta = -0 \cdot 2$ to $\theta = 0 \cdot 2$ rad,
       $= 0$ elsewhere.

$$\int_{-\pi}^{\pi} [D(\theta)]^2 \, d\theta = 2 \int_0^{0\cdot2} (1 - 10\theta + 25\theta^2) \, d\theta$$

$$= 2 \left(\theta - 5\theta^2 + \frac{25}{3} \theta^3 \right)_0^{0.2}$$

$$= 0\cdot4 - 0\cdot4 + 0\cdot133 = 0\cdot133.$$

Also

$$D(0) = 1$$

Therefore from eqn. (6.44),

$$\text{D.F.} = 2\pi/0\cdot133 = 15\pi.$$

(b) $\qquad [D(\theta)]^2 = 1 - \theta^2$ over the range $\theta = -1$ to $\theta = +1$ rad,

$\qquad\qquad\qquad = 0$ elsewhere.

$$\int_{-\pi}^{\pi} [D(\theta)]^2 \, d\theta = 2 \int_0^1 (1 - \theta^2) \, d\theta = 2 \left(\theta - \frac{1}{3} \theta^3 \right)_0^1$$

$$= 2 - \frac{2}{3} = \frac{4}{3}.$$

Also

$$D(0) = 1$$

Therefore

$$\text{D.F.} = 2\pi \, . \, 3/4 = 3\pi/2.$$

Neither pattern can be closely realized in practice, since discontinuities indicate the need for an infinitely long array.

6.6. $l \gg \lambda$, therefore

$$\sin \theta \simeq \theta \simeq \frac{\lambda}{2\pi} K.$$

So we can write D.F. from eqn. (6.44) as

$$\text{D.F.} = \frac{[D(0)]^2}{\dfrac{\lambda}{4\pi^2} \displaystyle\int_{-\infty}^{\infty} [D(K)]^2 \, dK}.$$

Inspection will quickly show that the integration of $[D(K)]^2$ is difficult. But we can use eqn. (6.49) to make the problem simple.

$$\frac{1}{2\pi} \int_{-\infty}^{\infty} [D(K)]^2 \, dK = \int_{-\infty}^{\infty} [T(r)]^2 \, dr = \frac{4}{l^2} \int_{-l/2}^{+l/2} \cos^2 (\pi r/l) \, dr = 2/l.$$

$$\int_{-\infty}^{\infty} T(r) \, . \, dr = \frac{2}{l} \int_{-l/2}^{+l/2} \cos (\pi r/l) \, dr = 4/\pi.$$

Eqn. (6.11) gives

$$D(0) = 4/\pi \text{ so that } [D(0)]^2 = 16/\pi^2$$

Therefore

$$\text{D.I.} = 10 \log_{10} \frac{16/\pi^2}{\lambda/4\pi^2 \cdot 4\pi/l} = 10 \log_{10} (16l/\pi\lambda)$$

and

$$\text{N.F.} = 10 \log_{10} l \cdot (2/l)/(16/\pi^2) = 10 \log_{10} (\pi^2/8).$$

6.7. Extending eqn. (6.44) to the case of a rectangular transducer

$$\text{D.F.} = \frac{[D(0,0)]^2}{\dfrac{1}{4\pi} \displaystyle\int_{-\pi}^{+\pi} \int_{-\pi/2}^{+\pi/2} [D(\theta, \o)]^2 \sin\theta \, d\theta \, d\o}.$$

For a linear array

$$D(\theta) = \frac{\sin x}{x} \quad \text{where } x = \pi l_1 \sin \theta/\lambda,$$

$$D(\phi) = \frac{\sin y}{y} \quad \text{where } y = \pi l_2 \sin \phi/\lambda,$$

thus

$$\text{D.F.} = \frac{4\pi}{\displaystyle\int_{-\pi}^{+\pi} \int_{-\pi/2}^{+\pi/2} \frac{\sin^2 x}{x^2} \frac{\sin^2 y}{y^2} \sin\theta \, d\theta \, d\o}$$

But

$$dx = \frac{\pi l_1 \cos\theta \, d\theta}{\lambda} \simeq \pi l_1 \, d\theta/\lambda$$

$$dy = \frac{\pi l_2 \cos\phi \, d\phi}{\lambda} \simeq \pi l_2 \, d\phi/\lambda$$

since θ and $\o$ are small when l_1, $l_2 \ll \lambda$. Also as the maximum response is centred on $+\pi/2$ direction, $\sin\theta \to 1$.

Thus

$$\text{D.F.} = \frac{4\pi}{\dfrac{\lambda^2}{\pi^2 l_1 l_2} \displaystyle\iint \frac{\sin^2 x}{x^2} \frac{\sin^2 y}{y^2} \, dx \, dy}.$$

Since $(\sin^2 x)/x^2$ falls off rapidly with increasing x the limit may be taken as $\pm \infty$.
Now

$$\int_{-\infty}^{+\infty} \frac{\sin^2 x}{x^2} \, dx = \pi.$$

Thus

$$\text{D.F.} = \frac{4\pi}{\lambda^2 \pi^2/\pi^2 l_1 l_2} = \frac{4\pi l_1 l_2}{\lambda^2}.$$

6.8. The range resolution $L = c\tau$ where τ is the transmitted pulse length. If $L = 5$ m, $\tau < 3 \cdot 3$ msec, say 3 msec. The receiver bandwidth must be at least of the order of c/L, i.e. 300 c/s, say 500 c/s.

The carrier frequency must be low to minimize absorption but high enough to ensure that (a) the dimensions of the transducer are conveniently realizable and (b) that the sea-state noise is not excessive.

Assuming that the transducer will be a linear array, reference to Fig. 6.3 shows that the 3-dB point of the response occurs at $Kl/2 = 0.44\,\pi$. The 3-dB beamwidth of such an array is therefore $2\sin^{-1} 0.44\,\lambda/l$. If this is to be $1°$ in the horizontal plane the ratio of λ/l is fixed and its absolute value is determined by the need to keep l to some convenient size. If 100 cm is taken as the maximum allowable value of the horizontal length the corresponding frequency is 75 kc/s. Reference to Fig. 3.5 shows that the attenuation coefficient at this frequency is 0.02 dB/m which would impose an attenuation loss of 80 dB over a total signal path of 4000 m, which is acceptable.

Assuming spherical spreading, the spreading loss is given by

$$40 \log_{10} 2000 = 132 \text{ dB}$$

making the propagation loss 212 dB. It will be assumed that a vertical angular resolution of $10°$ will suffice; thus the vertical length $= 10$ cm.

The directivity factor of the chosen transducer is given by $4\pi l_1 l_2/\lambda^2$ (see problem 6.7) and the corresponding directivity index is given by

$$\text{D.I.} = 10 \log_{10} \text{D.F.} = 35 \text{ dB.}$$

If the noise intensity is -160 dB relative to 1 W/m²n a 1 c/sband, the total noise intensity is

$$-160 + 10 \log_{10} 500 = -133 \text{ dB relative to } 1 \text{ W/m}^2,$$

This noise intensity will be reduced by the directivity of the receiving transducer. Following the usual practice of using the transmitting transducer for reception also the effective noise intensity level at the receiver is then

$$-135 - 35 = -168 \text{ dB relative to } 1\text{W/m}^2,$$

The target strength of a 2 m radius sphere $= 20 \log_{10} 1 = 0$ dB,

It is wise to allow a recognition differential at the receiver of 10 dB. Now the propagation loss must not exceed the difference between the required intensity level at the receiver and the source intensity level therefore the propagation loss = source level + target strength − received noise level − recognition differential

i.e.

$$212 = I_S - (-168 + 10).$$
$$\text{where } I_S - 10 \log_{10} W + \text{D.I} - 10 \log_{10} 4\pi$$
$$W = 30\text{dB relative to } 1 \text{ W}$$
$$= 1 \text{ kw}$$

A suitable specification is:

Frequency $= 75$ kc/s.
Transmitted power $= 10^3$ W.
Transducer linear dimension $= 100$ cm $\times$ 10 cm.
Pulse duration $= 3$ msecs.
Bandwidth $= 500$ c/s.

Index